A Million Cars For A Billion People

The story of India's 'millionaire' cars,
~ from the iconic Maruti 800
to what went horribly wrong with the Tata Nano

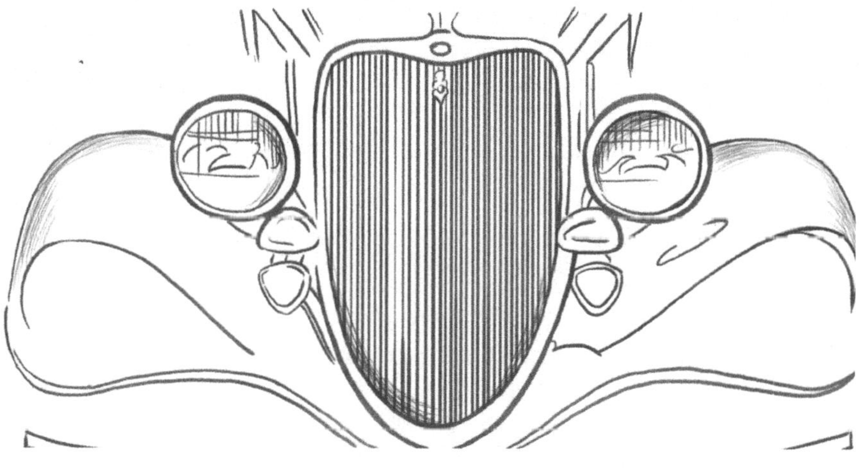

GAUTAM SEN

PLATINUM PRESS

ISBN 978-93-52013-53-1

Editing: Suchita Vemuri
Layouts: Ajay Shah
Cover Design: Fravashi Aga
Printing: Repro India Ltd

Published in India 2014 by
PLATINUM PRESS
An imprint of LEADSTART PUBLISHING PVT LTD
Trade Centre, Level 1, Bandra Kurla Complex
Bandra (E), Mumbai 400 051, INDIA
T + 91 22 40700804 **F** +91 22 40700800
E info@leadstartcorp.com **W** www.leadstartcorp.com
US Office Axis Corp, 7845 E Oakbrook Circle, Madison, WI 53717, USA

For Anne and Milan,
for bearing with me
and my late hours
once again...

About the Author

Gautam Sen graduated in economics before doing a post-graduate degree in Business Management. He began his automotive career in 1984, with Maruti Udyog. Acknowledged as a pioneer in Indian automotive journalism, Gautam became the Founder-Editor of India's first newsstand car magazine, Indian Auto, in 1986; and launched Auto India, India's bestselling car magazine ever, in 1993; followed by the Indian editions of auto motor und sport (2001) and BBC Top Gear (2005). Since 2007, he has been back in the driver's seat at Auto India. Gautam has also been involved in product development and consulted with Hero Motors, Hindustan Motors, San Motors, Tata Motors, TVS-Suzuki and several other prominent Indian, French and Italian automobile companies.

Living between Paris and Mumbai, Gautam has also authored, The Maharajas & Their Magnificent Motor Cars, The Car Design Book and Rolls-Royce 17X A Fabulous Destiny. He can be reached at gautam@gautamsen.com

Contents

Preface

A million cars for a billion people does not really add up to much. In mathematical terms it works out to a miniscule 0.01 per cent. So, ascribing a car the honour of being a 'people's car' simply because it has sold a million is perhaps stretching a point. Nonetheless, the handful of car models that have been 'millionaires' (*sic*), have indeed been engines of growth for India's automotive industry and the nation's economy.

But the growth happened mainly in the last three decades — against 14,688 cars sold in 1950, 2.65 million cars, SUVs and multi-utility vehicles found buyers in India in 2012. The Indian car market grew at an average seven per cent between 1950 and 1980. It was only in the 1980s — specifically with the introduction of the Maruti 800 mid-way through the decade — that the growth rate spiraled upwards, touching 56 per cent per year at one point. Among hundreds of new car models introduced over the last three decades, only a handful have been real vanguards of growth for the industry by setting the sales charts afire: Maruti 800, Tata Indica, Hyundai Santro and Maruti Alto. These cars have brought tremendous change in an India that has come a long way from the socialistic straightjacket of a few decades ago.

Whether these changes will be as impactful on India as the Ford Model T was on the USA, remains to be seen. The Model T was key in effecting the greatest and most rapid change in the lives of ordinary people in American history, transforming the automobile from being just a plaything or a form of sport for the wealthy to a product of use for the common man. But in India, in 2009, just 18 people in every 1,000 had a car, compared with the developed countries' average of well over 500 cars per 1,000 inhabitants. There is still a very long way to go in fulfilling the human desire — rightfully that of every Indian — for mobility, travel, freedom and most significantly, status.

Like the Model T, cars like the Volkswagen Beetle in Germany, Citroen 2CV in France, Fiat 500 in Italy and Austin 7 at first, later followed by the Mini in the UK, all became cars for the people, for the middle class, for all those seeking a means for easy travel and transportation of goods and services. All these cars had one thing in common: they became *the* engines of growth for the very important auto (and related) industries of USA, France, Britain, Germany and Italy and even, in the process, convinced leaders of many developing nations that what was needed for economic growth in every country was, at the least, a 'people's car'.

It took more than three decades in India's economic history for this idea to germinate and when it finally did, it happened thanks to essentially filial *raisons d'êtres* rather than cold economic reasoning. But the Maruti 800 that emerged from these was a true people's car, a car truly for the people. Tata Indica, Hyundai Santro and Maruti Alto have also been proper people's cars but evidently none could parallel the level of impact the Ford Model T had on America. With sales of over 15 million in less than two decades, the Model T drove away the horse altogether, causing a revolution in agriculture. It became the mainstay of the American economy and stimulus to urbanization, forever transforming the landscape of the countryside.

This has yet to happen in India. The Tata Nano, which was designed more specifically than any other to be the truest and ultimate people's car for India, was expected to revolutionise the automotive industry and evoke significant changes in the lives of both India's burgeoning middle class and those of all emerging nations. But despite being lauded by most as the 'right product at the right time' and 'the most significant new car since the Ford Model T was introduced 100 years ago', why did the Nano fail? What went wrong, given the unprecedented acclaim and praise? Is it really all over for the Nano? Or is there still some hope for India's 'little wonder'?

This book attempts to answer all the questions, concerns and doubts while clarifying the many misconceptions and myths that have gathered momentum over the years about India's automotive history and its mission to bring to the people a true 'people's car'. The very first cars that came to India and marked the early beginnings of the industry; the nascent history of the automobile across Germany, France, US, UK, Italy and even Japan and South Korea; all this is narrated from the viewpoint of the quest for the ultimate 'people's car'.

Chapter I
Engineering the Automobile Vision

The French will tell you that it was a Frenchman, Nicolas-Joseph Cugnot, who invented the automobile when his three-wheeled *fardier à vapeur* — a steam-powered dray wagon — made its first run in 1769. The following year, Cugnot developed a 'passenger' version on which four people could sit and be ferried around at about 4km per hour. Yes, it was quicker to walk.

A replica of Cugnot's fardier à vapeur *– the world's first self-propelled vehicle.*
© Gautam Sen

It took another century before 'proper' steam cars were actually being made and sold in some numbers. And it was another Frenchman, Amédee Bollée Sr., who started making very advanced steam cars from 1873. One of them, La Mancelle, built in 1878, had a front-mounted engine, shaft drive to the differential, chain drive to the rear wheels, steering wheel on a vertical shaft and the driver's seat behind the engine.

Now that sounds like a car, right? La Mancelle was perhaps the first automobile to be put in series production, with about 50 made over the next few years.

But the car, as we have known it for the best part of the last century, was the work of two Germans, living not too far from each other, but quite unaware of the other's invention and activities. Karl Benz and Gottlieb Daimler were not only the progenitors of the automobile era, but also carried the names that eventually merged to became one of the greatest marques of the automobile age. Yet, in all likelihood, they never met!

Benz, who lived in Mannheim, an old city in southwestern Germany, was an energetic technician. He established a production facility for gas engines in 1883 named Benz & Cie. Daimler, who lived near Cannstatt, near Stuttgart, breathed life into the first lightweight, high-speed petrol engine, working together with his colleague, Wilhelm Maybach, who would later go on to found his own top-end marque, Maybach. Daimler and Maybach transformed a long-held vision into a working reality. Though a French engineer Edouard Delamare-Deboutteville did build a single-cylinder four-stroke engine that was very advanced for its time, before Daimler and Benz, it is not certain whether Delamare-Deboutteville ever built a car. Daimler's universal four-stroke power source — registered with the German patent office in 1885 — was good for fitting in coaches, railway carriages, boats, ships, pumps, generators and, of course, the automobile.

Daimler first fitted the engine in a two-wheeler and his son, Adolf, undertook the first public test-ride on November 10, 1885. This was the world's first-ever motorcycle, but with outrigger wheels like those you see on learner bicycles for children. The *Reitrad*, or riding car in German, was registered with the patent office on August 29, 1885.

But patent no. 37435, granted to Karl Benz on January 10, 1886 for his Patent Motor Wagon, is the one that many recognise as 'the true' birth certificate of the automobile. The first public outing, though, of Benz's Motor Wagon was on July 03, 1886. And it was the remarkable journey undertaken by Karl's wife Bertha Benz, in August 1888, that really established the automobile as a serious alternative to the horse and horse-drawn carriage. Frau Benz, with sons Eugen and Richard, drove 180 kilometres from Mannheim to Pforzheim and back in a Benz Motor Wagon, proving that the automobile was more than just a plaything for dilettantes.

A replica of the Benz Motor Wagon, showing how basic it was. © Gautam Sen

Interestingly, the Benz Motor Wagon was a three-wheeler. After about 25 Motor Wagons had been made, Benz launched the Victoria B in 1893 — his first four-wheeled motor car. 1884 saw the debut of the Benz Velo, an attractively priced, lightweight car for two — the first ever 'people's car'! It was the Velo that ensured Benz's breakthrough to higher sales figures, with production of around 1,200 samples altogether. Thus, the Benz Velo became the first automobile to see large-scale production. By 1895, Karl Benz had made 1,132 vehicles of all types and, at the turn of the century, Benz had become the world's leading automobile manufacturer.

Yet, it wasn't Germany that was leading car manufacturing activity — that prerogative was with the French. By 1900, France was the world's biggest car-making nation. The pioneer among the French car-makers was Panhard et Levassor. Founded in 1890, Panhard et Levassor produced its first car fitted with a Daimler engine. The four-wheeled 'Systeme' model that the company produced in 1895 was possibly the first car fitted with a sliding gear transmission. It was also the first car with a front-mounted engine with the drive going to the rear, to drive the rear wheels — an installation that became known as the FR (front-rear, or engine at Front with drive to the Rear) layout; a configuration still used by most big cars produced today.

The other pioneer among French car-makers was De Dion-Bouton. Founded in 1883 to make steam cars, De Dion-Bouton turned to internal combustion engine cars by 1889 and, for some years, was the world's biggest automobile manufacturer.

The Peugeot Bébé, an early people's car, that was all of 2.7 metres long and weighed just 350 kilograms. © Gautam Sen

Another noteworthy early French car-maker was Peugeot. Though Armand Peugeot's first car was a steam-driven car built in collaboration with Leon Serpollet (who went on to make a series of very advanced steam cars), the former soon moved to making a petrol-engine car with a Daimler-licensed, but Panhard-produced engine. By 1896, Automobiles Peugeot was making its own engines and, by 1899, had sold 300 of the total 1,200 cars made in France. The numbers went up so rapidly that, by 1903, Peugeot made almost half of all French cars — a significant number, as more than half the 63,000 cars produced globally in 1903 were manufactured in France by some 30-odd car-makers, including Peugeot.

Renault was the other important French car manufacturer. Though it started out a little late — 1899, in fact — it built a very fine reputation quickly, specifically for its tiny Voiturette (small car in French), which racked up impressive sales numbers. Yet the price of the smallest Renaults available then was 3,000 French francs, equivalent to a factory worker's wages for 10 years! Far from what could be called a real 'people's car'.

Louis Renault on his Voiturette, another early people's car.
Renault Corporate Communications/All Rights Reserved

By 1904, the French lead in car manufacture was overtaken by the Americans. Leading the American 'charge' was Olds Motor Works, a pioneer in the country's burgeoning car industry. Founded by Ransom E Olds in 1897, Oldsmobile, as it was colloquially referred to initially, became America's top selling brand with the introduction of the model called Curved Dash, which is today acknowledged as the first mass-produced car ever. (Though Peugeot was also producing many cars, it made several different models with considerable variation in specifications.)

But it was the pioneering efforts of another American that changed the world of car-making altogether. Born to a farming family in Michigan on July 30, 1863, Henry Ford was the eldest of six children of William and Mary Ford. Grandson of Irish immigrants who had arrived in America in 1847, Henry Ford didn't much care for farm work. Instead, it was mechanical and engineering intricacies that fascinated him.

Intelligent and curious by nature, Ford was excited by the enormous growth of industry in Detroit. An avid experimenter, Ford once, to prove the power of steam, stopped the steam of a kettle full of boiling water. It exploded! At 15, Ford built his first steam engine. Once he was a little older, his father allowed him to rework the tools at the farm, with his mother providing mending needles and corset holders, with which Henry 'invented' the tools to make the repairs.

The decisive event in the life of Henry Ford came about in 1876, when the 13-year-old Henry, travelling with his father in a wagon, came across a steam engine, operating under its own power. Henry jumped out of the wagon and ran to the driver of the vehicle to question him. Not only did the driver explain the workings of the machine, he even let Henry operate it. Having seen a vehicle that could propel itself, the thought of making one himself haunted Henry's imagination for years.

Although he dreamt of going to Detroit and working in the factories, Henry remained on the farm to support the family till the age of 17. Then, with the blessings of his father, he went to Detroit, becoming an apprentice mechanic in the Detroit stores of James F. Flower and Bros. After his apprenticeship in 1882, he spent a year adjusting and repairing steam engines for Westinghouse in Southern Michigan.

At a party in 1885, he met Clara Jane Bryant, the daughter of a farmer from Wayne County. They got married on April 13, 1888 and had a child, who they named Edsel, born on November 1, 1893.

Even as he was raising his little family, Henry hadn't abandoned his dream of designing a 'horseless carriage'. Whenever he had a free moment, he read books on gas engines and experimented in his workshop. In 1891, he and Clara moved to Detroit and Henry began working for Edison Illuminating Company, the company founded by genius inventor Thomas Edison. In December 1893, he became chief engineer for the company, heading the maintenance of the electric supply to the city. As the hours were irregular, it allowed him the time to experiment.

In a small workshop behind his home, Henry Ford developed the final details of his gasoline-powered car. After more than two years of trials, Henry Ford, at age 32, had completed his first experimental automobile. He named it Quadricycle, as the car rolled on four bicycle wheels. This machine frightened the horses and was the cause of some protests in the neighbourhood. But it worked. The success of this small vehicle fanned the ambitions of Henry Ford.

Thomas Edison was impressed and encouraged Ford to develop a second vehicle, which was ready by 1898. Backed by wealthy investor William Murphy, Ford left Edison to start the Detroit Automobile Company in 1899. But the enterprise was not very successful and the company was dissolved in 1901. Ford again teamed up with Murphy to launch the Henry Ford Company, which he left soon after, as Murphy brought in another very able engineer, Henry Leland, and the Henry Ford Company was renamed Cadillac Automobile Company (after Antoine Laumet de La Mothe, sieur de Cadillac, the French founder of the city of Detroit).

It was 'third time lucky' for Henry Ford when his third automotive venture, the Ford Motor Company was incorporated on June 16, 1903, with a capital of $28,000. The first car that rolled out of Ford Motor Company was the Model A, of which 1,750 cars were made between 1903 and 1904, all red. The sardonic offer for the Model A was that the buyer could have it in any colour as long as it was painted red!

In 1904 a bigger car, the Model B followed. The Model C was the Model A with a more modern look. By 1908, Ford had got to his ninth model — the Model T.

Engineering the Automobile Vision

Introduced on October 1, 1908, the Ford Model T changed the world. By the time the production of the Model T came to an end — on May 26, 1927 — more than 15 million of the cars had been produced (15,458,781 to be exact!) making Ford not only the number one carmaker in the world, but changing the landscape of America, changing the way Americans lived — as one generation transited from the horse to the motor car — and giving a new meaning to the 'American Dream'.

So much so that by 1929, just before the Great Depression, of the 32 million automobiles in use across the globe, almost 90 per cent had been produced (and most of them were being used) in the USA! A kind of 'domination' that was not to be seen ever again. And most of that was thanks to Ford Motor Company and the Ford Model T.

Ford Model T – the first true people's car. Courtesy Ford Motor

The Model T was not revolutionary per se. It was a logical evolution of the cars that Henry Ford had been developing. After the Model A, the Model B, the C, the F and the K, the Ford Model N, launched in 1906, was an important development. Other than the fact that the Ford Model N replaced the A, C and F series, it also sold in significant numbers, with some 7,000 produced between 1906 and 1908, as it was small, nimble, and rather reasonably priced at US$500 then, which would be about US$12,000 (or Rs 6.5 lakh) in today's monetary value terms. The most noteworthy achievement about the Model N was that Ford had managed to bring down the price to half that of the Model F by intelligent value engineering.

The Model N was a precursor to the Model T, as the car featured an in-line four-cylinder engine (unlike the twin cylinders of the A and F series), located at the front and driving the rear wheels with a shaft, like the Syteme Panhard. Though it was very basic transport, the Model N quickly built up a reputation for being rugged and practical and Ford became known as a manufacturer of dependable, inexpensive cars for a broad market.

The Model R that followed was an upgraded version of the Model N that was more expensive (at US$750); yet, in its short production run from April to October 1907, some 2,500 of the R got sold.

The direct predecessor of the Model T though, was the Model S. At US$700, it was slotted between the N and the R but with more elaborate bodywork that included a cowl-like cockpit and improved fenders, complete with running board and splash screen.

The Model T was a natural culmination of the preceding models from Ford and other automobiles, as they were evolving into reliable transportation for the middle class and the wealthy in the USA. It wasn't the smallest or cheapest, but best suited to the needs of the American hinterland then. Powered by a 20-horsepower (hp), four-cylinder engine that displaced 2896cc, the Model T was capable of a max speed of about 45mph (72km/h), which was fairly impressive for its time. At about 3.4 metres (depending on the version) the Model T was not that small either. There was enough space for four very healthy Americans to ride comfortably in its tourer (a four-door saloon with conventional seating and removable side screens and folding top) or sedan body style. But most importantly, the Model T had very impressive ground clearance and, with the big wheels and amazing suspension, thanks to its transverse leaf springs, could easily negotiate the rough and rutted dirt tracks of deep America. It was the kind of car that America needed then.

At its launch, the Model T wasn't cheap either: US$850 for the base version was like paying almost US$21,000 (Rs11 lakh) for a car today. But it was a wee bit cheaper than its most direct rival then, Buick's Model 10, which, at US$900, was the slightly pricier volume leader of the Buick range at that point of time. Buick's Model 10 accounted for about half of the carmaker's volume then, with as many as 8,820 Model 10s selling in 1908, as compared to 4,641 Model 10s sold in 1907.

Henry Ford and his Model T. Courtesy Ford Motor

Buick's Model 10 was also perhaps a better car than Ford's Model T. Other than the slightly smaller engine developing more power, the Model 10 had a more sophisticated suspension system, a rear differential housing that was stronger than the T's and three pedals instead of the Ford's two-pedal initial set-up (which was later changed to three-pedals). But then Buick owner William C Durant went on an acquisition binge as he started to build up General Motors, starting 1908. By 1910, Durant was out of GM, as the General needed financial restructuring.

On the other hand, Henry Ford concentrated on making his one much perfected car: the Model T. In 1909, the first full year of production of the Model T, Ford sold 10,660 cars, as long as they were red, green or grey in colour. The numbers went up to 19,050 Model Ts sold in 1910, to 34,858 in 1911 and 68,773 in 1912. Steady growth, nothing sensational.

But all the while, Henry Ford was tinkering and perfecting his manufacturing set-up, working at reducing the time, effort and cost needed to make his Model Ts. Ford broke up the making of the Model T into 84 discrete steps and he trained his workers to do just one step well. He also hired motion-study expert Frederick Taylor to make the activities of the workers more efficient. And then on December 01, 1913, Ford introduced the first moving assembly line for the mass production of the automobile.

The famous Ford assembly line. Courtesy Ford Motor

Production of the Model T jumped to 170,211 in 1913. In 1914, 13,000 workers at Ford made 260,720 Ts as compared to 286,770 cars made by the rest of the industry with as many as 66,350 workers. Despite paying his workers significantly more than the industry average, the cost of making the Model T dropped. And, instead of pocketing the profits, Ford dropped the price of the Model T to as low as US$260 by 1925. Sales of the Model T skyrocketed, topping a million sold in 1921, two million in 1923. Ford sold more than half the cars in the global automobile industry during 1918-1919 and 1921-1925.

And, along with mass production came the adage that you could have your Model T in any colour as long as it was black. And though there were several different variants to choose from — tourer, roadster, coupé, two and four-door sedan, pick-up, ambulance and also as chassis only — change was essentially limited to details. By the mid-1920s, the rest of the industry had progressed beyond the Ford Model T to more modern and better cars and before Henry Ford knew what was happening, sales of the Model T skydived from 1.6 million sold in 1926 to just 450,415 in 1927. In October 1927 the plug was finally pulled on the Model T.

With the demise of the Model T, Ford's dominant position suffered and Henry's company slipped behind General Motors, a situation that hasn't changed since the late 1920s. Yet the impact of the Model T and Henry Ford's entrepreneurial philosophy on 20th century industry has been immense. Not only did mass production and assembly line manufacturing become the bywords for modern industry but the possibility of making cars accessible to the masses became an integral part of industrial and economic growth for every major nation since then.

All the people's cars...

According to the vision of Henry Ford, the automobile had become a product of use for the common man, rather than being only accessible to rich people, who were the only ones able to travel freely in the country. The Model T was the key instrument of the greatest and the most rapid change in the lives of ordinary people in American history. And all this change happened in less than two decades.

From being a plaything, a toy or a form of sport for the wealthy, the automobile became a tool for everyday use by everyday people, a product of convenience, for mobility and accessibility. The automobile became a car for the people, for the middle class, seeking a means for easy travelling and the transporting of goods and services.

The horse disappeared so quickly that the transfer of land for hay and crops caused a revolution in agriculture. The car became the mainstay of the American economy and a stimulus to urbanisation — cities were formed, suburbs came into existence, the emergence of residential development and the making of major highways happened. As automobiles needed frequent maintenance, repairs garages and storefront shops were opened by mechanics, blacksmiths, bicycle repairers and others. Car dealerships and fuel stations came up as did places for parking, cleaning and other services for motorists.

In 1922, at the height of the Model T success story, Henry Ford (in collaboration with writer Samuel Crowther) published a book called 'My Life and Work'. It wasn't an autobiography as one might be led to think, but a manual of Ford's business philosophy. The book describes how Ford introduced the assembly line, reduced working hours, institutionalised the concept of minimum wages, the five day week, and so on, a philosophy that was soon called 'Fordism'. A "virtuous economic and social compromise" that came by with mass manufacturing and the improvement in wages, Henry Ford wanted his workers to be well paid so that they in turn could buy the cars that they produced. Fordism also found favour with British economist John Maynard Keynes, who echoed the idea by saying that, "Steadily increasing wages at the rate of productivity growth will ensure that the opportunities also offered to businesses will grow at the same rate and will therefore avoid overproduction."

Of course, with the oil crisis of the early 1970s and the changing world order since then, Fordism is now seen as a concept of the past, with several scholars propounding post-Fordism, also referred to as 'Flexibilism', which talks of small batch production, economies of scope, specialised jobs and the increasing importance of service and white-collar workers.

But Ford's famous quote, "I will build a car for the great multitude. It will be large enough for the family, but small enough for the individual to run and care for," has remained a source of inspiration to generations of industrialists and car-makers since.

And "it will be so low in price that no man making a good salary will be unable to own one." The maxim has driven entrepreneurs, engineers, inventors, dictators and charlatans to champion the cause of 'the people's car' since the Ford Model T.

Of the many who were inspired by Henry Ford and his Model T, the two most prominent, in the aftermath of the First World War, were Englishman Herbert Austin and Frenchman André Citroën.

André Citroën was born in Paris on February 05, 1878, the fifth and last child of Levie Citroen, a Jewish diamond merchant from Holland, who had migrated to Paris in 1873, and Masza Kleinman, a Polish Jew born in Warsaw. The name Citroën has an interesting history. André's great grandfather, Jacob was a citrus merchant in Holland. In 1810, when Napoleon annexed the Kingdom of Holland, the people were subjected to the Napoleonic code and had to choose a name for their identification. Jacob chose to call himself Limoenman (the lemon man) a nickname that his clients had given him. But his son, Roelof didn't care for Limoenman, and so instead opted to change that to a more French sounding Citroen (citron is lemon in French). The change in name didn't seem to help business though and Roelof Citroen moved to the gem trade, which was a growing sector in the 19th century.

In 1884, André's father Levie Citroen committed suicide by jumping out of the window when André was just six years old. Levie Citroen had gone into a high-risk speculative venture in connection with a diamond mine in South Africa, and though the family was not completely ruined, the bankruptcy weighed heavily on Levie's shoulders. André was then raised by his mother who took over her husband's business of trading in diamonds and pearls. The Citroen family received a proper French education, so that the children could grow up as French.

In school, André proved to be a brilliant student, with a special talent for science and technology. It was also in primary school that the umlaut to the 'e' in Citroen — making it Citroën — was added, by either a teacher or the headmaster. André was just nine years old when he discovered the work of avant-garde visionary Jules Verne, and that inspired him for the rest of his life. The construction of the Eiffel Tower for the Universal Exposition of 1889 also encouraged him to become an engineer and begin participating in the future industrial challenges of the 20th century. In October 1898, André Citroën joined the Grande École Polytechnique, France's most prestigious engineering school, which also provided an opportunity for Citroën to be exposed to a vast network of very influential people.

In 1900, during the Easter holidays, when André Citroën was visiting family in Poland, he came across the client of his uncle (mother's brother) who had developed a process of making herringbone gear teeth cut into a V, from a Russian patent. André saw an opportunity in branching out as an independent manufacturer and so he promptly secured the licence to produce and market them. In later years André adopted the inverted double V or 'doubles chevrons' as the French call them, as his personal trademark and eventually the trademark of the marque that he founded.

For the gears to work, the teeth had to be perfectly machined. The small Polish firm couldn't afford such precision manufacturing, and neither could any of the larger manufacturing companies in Europe. So André Citroën turned to America, where the manufacturers of machine tools had the technological know-how. But before he could begin production of his gears, Citroën had to finish his last year of study at the Polytechnique and his years of military service which was mandatory at that time, spending two years in the 31st Artillery Regiment of Le Mans, as a sub-lieutenant.

On completion of his tour of duty, André Citroën landed the job of running pioneer French car-maker Mors in 1906. Founded by brothers Louis and Emile Mors, the car-maker was in financial difficulties despite its fame in establishing speed records and beating most of the others in racing. Citroën reorganised the management, upgraded the cars, in the process creating new models and doubling the production of the brand in 10 years. In 1908, Mors made 300 cars, but by the 1913 the number had gone up to 800 units.

Early on, it became obvious that André Citroën had a genius for scouting talent and organising. He was neither an inventor, nor a true engineering wizard. His passion was for the business of "manufacture and distribution on a large scale." In 1912, as part of his work for Mors, André Citroën made his first trip to the United States, a trip that profoundly affected his life and career. He visited the Ford factory at Dearborn, near Detroit, which was in the process of applying Taylorism and Fordism. Citroën became an immediate fan of the 'democratisation' of capital goods and industrial mass production at low cost. Citroën made the most of that visit, absorbing and learning the methods used in American factories.

Finally, it was in 1912, at the age of 35, that Citroën teamed up with André Boas and Paul Hinstin, both of who he knew from school, to invest a large part of his inheritance to start the Citroën, Hinstin et Cie gear-manufacturing company, specifically making the herringbone V, with a dozen workers in the 10th arrondissemènt (district) of Paris. Soon he moved to a bigger facility and the gear-making company had, as its logo, Citroën's favourite — the double inverted V. André Citroën founded his reputation and fortune in the production of these gears, followed by artillery shells for the French army during the First World War, where he applied the concept of Fordism to mass manufacture shells in an assembly-line like facility. By the time the war was over, Citroën had produced 23 million shells!

When the Armistice was signed in 1918, the Citroën factory, with its 13,000 workers came to a standstill. But Citroën, in anticipation of this, had begun reinvesting his profits into a new enterprise. The automobile was chosen to replace the production of ammunition. André Citroën converted his weapons factory to one ready to make cars within four months; in the process, he absorbed the Mors brand. And sure enough, the Citroën brand took on the double chevron emblem in its historic inverted V shape.

A key point that Citroën realised during his trip to the United States was the need to produce an automobile whose market had to extend beyond the privileged classes, a product that would be both useful as well as a source of entertainment. André Citroën announced to the French press in January 1919 his intention to produce an automobile for the masses, a people's car.

Besides the fact that the future automobile had to sell for a target price of 7,250 French francs, or nearly half the cheapest car on the market at that point of time, the ad that accompanied the announcement was remarkable in automotive history. In fact, it must have been the first massive press campaign ever in the history of the automobile. The result was that Citroën had more than 16,000 serious enquiries in just two weeks.

Assisted by his trusted right-hand man, Georges-Marie Haardt, in marketing the product, André Citroën rolled out in 1919 the first model of the brand, the Citroën Type A, for which he had already recruited engineer Jules Salomon in 1916. The Citroën Type A became in 1920 the "first mass-produced European car" produced at a rate of 30 cars per day initially and then going up to 20,000 per year by 1921. The success was unprecedented, especially given the exceptionally low introductory price for the time, which, of course, was at the expense of much needed profits to repay the substantial sums borrowed for the necessary investments.

The Citroën Type A – Europe's first mass produced car. © Gautam Sen

In 1926, Citroën developed, with the help of American engineer Edward Gowan Budd, the Citroën B14, an "all-steel" bodied car that featured a completely closed body, a first in Europe. Huge stamping presses stamped out 400 panels every hour, which were then assembled by welding these to the rest of the body. In 1928, Citroën launched the C4 and followed that up with the C6. By 1929, with 102,891 cars produced, Citroën had become Europe's biggest car-maker, and the world's fourth biggest (after GM, Ford and Chrysler Corporation). Citroën alone accounted for a third of total French car production then, and a little less than half of the cars were exported. By then, Citroën had five factories and employed 32,000 workers.

But it was not just Fordism that took Citroën to such levels of success. André Citroën was a master at marketing innovation! When Citroën began manufacturing cars, it opened a special department that was mainly responsible for accumulating, from the dealer network, information about the clients, their names, addresses, and so on. A letter would then be sent to the customers who had visited the dealerships, thanking them for their visit and informing them of the model that would best suit them. Though this letter was printed, the signature and the person's name would be hand written, giving the impression that the recipient of the letter was receiving one personally from André Citroën! This was long before computers and emails. Thousands of letters were distributed throughout France every week, dramatically improving sales results. In fact, according to company officials, 15 per cent of sales were the direct result of this strategy.

And it was not just direct marketing but at media stunts too that André Citroën excelled. On October 04, 1922, at the opening of the 17th Mondial de l'automobile de Paris (the Paris motor show), a plane drew the Citroën letters in white smoke above the city's Arc de Triomphe monument. In 1925, André Citroën installed on the Eiffel Tower a huge illuminated advertisement, spelling out 'Citroën' vertically, with light bulbs, 250,000 of them, in six different colours! Then there's the story of how Charles Lindberg used the brightly lit Eiffel as a beacon to approach Paris' Le Bourget aerodrome after his epic flight across the Atlantic, aboard the famous aircraft Spirit of St. Louis. With a penchant for the dramatic, André Citroën invited the international aviation hero to a reception at the Citroën factory, once again hogging the limelight for his company.

But for all the flamboyance and marketing flair, the company had feet of clay. André Citroën was a brilliant visionary and a great expansionist, not an accountant. In fact, the firm had financial problems from its inception. Gains made during the production of munitions were not sufficient to cover investments, as 90 per cent of the profits were used to pay taxes. Sales of the cars definitely helped, but the rapid growth of the firm came with too many expenses.

In 1933, André Citroën brought in brilliant engineer André Lefebvre, who had worked at avant-garde car-maker Voisin and Renault before. He was asked to design and prepare the production of an all-new cutting-edge mid-sized car, to be presented to dealers on March 24, 1934. Revolutionary in more than one respect, the amazing Citroën Traction Avant featured front-wheel-drive and monocoque construction. But before the Traction Avant could get into series production to save Citroën, the company went into liquidation.

In 1935, at the age of 56, André Citroën had to sell his shares to tyre-maker Michelin, one of the suppliers to which it owed the most. And in July that same year Pierre Michelin took over. On February 18, 1936, André Citroën was admitted to the clinic Georges Bizet. Although operated in May, Citroën died on July 3, 1936, of stomach cancer, never living to see the Citroën that — a little over a decade later — would be the true people's car.

The story of Herbert Austin across the channel was much less dramatic and more predictable. Born on November 08, 1866, to farmers in a village called Little Missenden, in southeast England, Austin went to live in Australia with his uncle, when just 17 years old. There, he turned to the engineering profession and apprenticed in different industries, including machining, locomotives, machinery for mining gold, and so on. After he completed his training, he joined the art school of Hotham in Melbourne, where he developed his drawing skills. He met his wife, Helen Dron, in Australia, and they had three children. Living in Australia, Austin joined the Wolseley Sheep Shearing Machine Company, working with Frederick Wolseley himself.

When Wolseley decided to wind up operations in Australia and head back to the UK in 1893, Austin came along with him. In the meantime, Austin had also taken a fancy to the new-fangled toy, the automobile. He designed two prototypes, both three-wheelers, and then convinced the Wolseley Sheep Shearing Machine Company to add a motor car in their catalogue. The Wolseley Sheep and Shearing Machine Company became the Wolseley Tool and Motor Car Company in 1901, when it was taken over by Vickers, and Herbert Austin was made the managing director of this company.

Tired of working for others, Herbert Austin decided to venture forth on his own in 1905. He quickly found a disused factory in Longbridge, near Birmingham, and established the Austin Motor Company. By 1906, he was already testing the first Austin, a 25/30hp 5-litre four-cylinder engine with chain drive. Three years later, the range consisted of three four-cylinder cars, the 15hp, the 18/24hp and the 40hp and a six-cylinder 60hp model which was the basis for a racing version that finished in the 18th and 19th places at the Grand Prix de France that year. In 1910, a four-cylinder 1.6-litre car was developed for export, but was instead sold on the UK market a year later.

During the First World War, Austin, like Citroën, produced munitions for the British war effort; from employing 2,500 on the eve of the war, Austin was employing 22,000 by 1917. Also, when war ended, and Austin Motor Company decided to go back to car-manufacturing as its principal activity, Herbert Austin took the decision to concentrate and mass produce — like Ford — a single model, the Austin Twenty.

But the 3610cc four-cylinder Austin Twenty was way too expensive for post-war England and the car didn't take off. By 1921, Austin Motor Company was in serious financial difficulties and a receiver was appointed. Herbert Austin lost full control of the company and he had to accept two other directors — Ernest Payton, to look after finance, and Carl Engelbach, to supervise production.

Fortunately, the recovery plan hatched between Herbert Austin, Payton and Engelbach worked and in 1922 the Austin catalogue expanded to offer the Austin 12, a smaller 1661cc engined car. This car gave Austin some volume as it appealed to the middle class.

But the real ace up Austin's sleeve was the 7, launched in 1922.

With the economic devastation of the war, Herbert Austin realised that the automotive market had changed. The traditional clientele no longer had the same purchasing power, yet the masses were demanding the right to own a car. It was therefore urgent to provide a minimalist automobile. There were already several 'cycle-car' manufacturers, who used mechanical elements borrowed from motorcycles, producing cars that were essentially spindly four-wheeler motorcycles, and who were trying to meet this demand, but the lightness of the vehicles and their tandem seating arrangement (one behind the other) limited their appeal.

The Austin 7 – UK's answer to the Ford Model T. © Makarand Baokar

Austin made the right choice in designing, with the very young engineer Stanley Edge, a car with reduced proportions, but with the parts vital for reliability, along with providing a minimum of comfort. He chose a triangular steel frame profile, short for good rigidity, a car engine (not an engine from a motorcycle), an in-line four-cylinder side valve powerpack with a proper three-speed gearbox (along with reverse), plus relatively big 19-inch wheels, to make sure that ground clearance was significant.

To keep his project secret, Herbert Austin shut himself up with his engineer and draftsman in his country house while developing the little car. The production of the Austin 7 began in 1922, surrounded by general scepticism. However, given the speed at which the car took off, Austin had to adopt the assembly line system, like Ford and Citroën. This reorganisation led to a dramatic decline in prices, which, in turn, boosted sales. When launched, the Austin 7 cost just £165, but, by 1932, the prices had come down to £110 (about £6,000, or Rs5 lakh, at today's values).

The Austin 7 was the British answer to the Ford Model T, but in two-thirds scale. It too could seat four, but a rather slim four in a tight fit. Power was also from a four-cylinder engine, but it was just 696cc (747cc soon after) as compared to the Model T's 2896cc, and the engine developed just 10hp, which was sufficient to give the lightweight 360kg car a top whack of almost 100kph! Yes, it was actually quicker than the Model T. This reflected in the Austin 7's sporting successes too, which included a third place at the 1929 Tourist Trophy and the leader in the 500 miles endurance race at Brooklands, in the UK.

Herbert Austin was only too aware of the phenomenal market success of the Ford Model T, and firmly believed that a car that was smaller and lighter, but also more affordable, would work perfectly for a country like Britain, known for its narrow cart tracks and winding lanes lined with hedgerows. For inspiration, he may have been influenced by the Peugeot Quadrilette, which the French car-maker had exhibited the previous year. The Austin 7, like the Quadrilette, was disarmingly simple. The front of the 7 was suspended by a transverse leaf spring and the rear used quarter elliptic springs on the early cars; shock absorbers were considered unnecessary at the beginning. Yet the 7 had brakes on all four wheels, when many cars still had braking on the rear wheels only.

With the success of the 7, Austin had 37 per cent of the British automobile market by 1929. Staying true to its original purpose of affordable transportation, the 7 received many detailed improvements and, by 1939, when production ceased, almost 290,000 Austin 7s had been made, not counting those made under licence in various corners of the globe.

In the UK itself, the Austin 7 proved to be an inexpensive, yet efficient base for making special, exclusive cars. A pretty two-seater was made by coach-builders Avon Body Company of Warwick, which, incidentally, was the first design from the Jensen brothers, who went on make the Jensen sports cars.

In 1927, William Lyons, the co-founder of a motorcycle sidecar-making venture, the Swallow Sidecar, like the Jensen brothers, saw commercial potential in producing a sportier version of the Austin 7. Lyons commissioned the talented coach-builder Cyril Holland to produce a distinctive open tourer, which was called the Austin Seven Swallow. With its bright two-tone colour scheme and a design that you would normally see on more expensive cars of the time, together with its low price of just £175, the Swallow proved to be very popular.

In 1928, a saloon version followed. Approximately 3,500 bodies of various styles were produced until 1932 when Lyons started making complete cars under the SS brand, with the 1936 SS 100, branded a Jaguar. Such was the demand for the Austin Swallows that Lyons was forced to move in 1928 from Blackpool to new premises in Coventry. And it was, in part, the success of the Swallows that laid the foundations for what was to become, by 1945, Jaguar Cars.

It was not just Jaguar that owed its origins to the plebian Austin 7. Versions of the 7 were made under licence by Rosengart in France from 1928, by American Austin (later Bantam) from 1930, chassis and running gear were exported to Japan and Australia to have locally made bodies attached (in Australia they were bodied by Holden, which became a part of GM by 1931) and the 7 was also the car that Nissan used as the basis for its first passenger cars.

It was also in 1927 that German car-maker Fahrzeugfabrik Eisenach took out a licence to make the 7 in Germany, branded as the 3/15 DA-1. Fahrzeugfabrik Eisenach was a German bicycle and gun manufacturer that had diversified into making cars in 1898, when it took a licence from French carmaker Decauville, making it Germany's third car-maker, after Benz and Daimler. By 1903, it had introduced a car of its own design, branding the car Dixi, when the Decauville licence was revoked. During the First World War, Fahrzeugfabrik Eisenach produced trucks and guns and, in 1919, car production resumed once again; however, the company struggled in the post-war scenario. In 1927 it took the decision to produce the Austin 7 under licence in Germany.

In late 1928, Fahrzeugfabrik Eisenach was taken over by BMW, which was looking to get into the car-making business. After 9,307 Dixi 3/15 DA-1 had been made between 1927 and 1929, the car became the BMW 3/15, BMW's first car. The 3/15s were made till 1932, with 18,976 produced, before it was replaced by the BMW 3/20, a BMW-designed evolution of what was still basically the Austin 7. Of course, with subsequent models, BMW moved further and further away from the plain Jane 7, but the beginnings of what is today the world's largest luxury car manufacturer, incidentally, was in a rather plebian people's car!

The BMW 3/15, which began life as a Fahrzeugfabrik Eisenach Dixi.
© Gautam Sen

But as BMW moved further upmarket with its in-house developed products, the then leader of the nation had different ideas. A certain Adolf Hitler had a dream … to initiate a 'people's car', for his people.

Unlike the 'people's car' from the USA, France and England, Germany's 'people's car' had a very different history. Here was a car that was designed by an engineer who could not initially find a manufacturer. The war delayed the car's launch by seven long years. The customers who had initially 'subscribed', not only did not get their car, they all lost their money. Thus, it wasn't exactly a success at the beginning. It was only later in its career that the car really started selling.

Hitler's envisioned car for the people, the Volkswagen (German for 'people's car') had a most inauspicious genesis, but a certain genius, by the name of Ferdinand Porsche, put metal to the dream.

Born in 1875, the Austrian-German Porsche rapidly established a reputation as an excellent automotive engineer beginning with an electric vehicle, the Lohner, in 1897, followed by several other automobiles, including the amazing 140km/h Austro-Daimler Prince Henry from 1910 and, finally, the fabulous Mercedes-Benz 38/250 SS and SSK in 1928. In 1931, Porsche went independent and founded a consulting firm in Stuttgart, with his son Ferry. Porsche's intention was to specialise in developing both racing machines and widely affordable low-cost cars — two very different objectives. For the former, the engineer almost always had carte blanche to get the best performance. In the case of the latter, Porsche had to think of every possible savings on each component and at every stage of manufacturing.

Porsche presented his draft for a low-cost car to several manufacturers, but only Zündapp and NSU, two German motorcycle brands, were interested. The former lost interest as its motorcycles sales took off and the latter found the tooling costs too high. Contacted by the leaders of a young Soviet Union, in 1932, Porsche was offered full control over the Soviet automobile industry, but he declined.

It was finally in 1934 that Adolf Hitler ordered the German Manufacturers' Association — which was otherwise reluctant to get involved in a project that would affect its members' existing product range — to liaise with his favourite engineer. Porsche built three prototypes, working at his villa, and requested Mercedes-Benz to lend its factories for a pre-production run. Finally, in 1938, Hitler was able to unveil his people's car, the 'strength through joy' car, the KdF-Wagen.

Hitler declared that the car would be priced at just 990 Reichsmarks, or the equivalent of less than US$4,000, or Rs2 lakh, today. Porsche knew that it would be commercially impossible to do that. But there was no way that he could question Hitler. All the more so as the Reich was going to finance racing too, for which Porsche was already working on an extraordinary Grand Prix car, one powered by a supercharged V16, developing more than 500 horsepower!

In contrast, the people's car's specifications were at the other end of the spectrum. Ferdinand Porsche's choice of mechanical configurations for the Volkswagen was an 1131cc air-cooled flat-four engine at the rear, developing all of 25hp and driving the rear wheels — a radically different solution from that of the Austin, Citroën and other cars from that period.

But was it original? No, not really, as Czech engineer Hans Ledwinka, had already designed rear-engine cars for Czech car-maker Tatra, starting with the T77 in 1934. Ledwinka apparently discussed his ideas with Porsche, who used many Tatra design features in the Kdf-Wagen. This is particularly evident when compared with the smaller Tatra T97 model, which had a rear-mounted, air-cooled, flat-four engine and an aerodynamic body styling. Incidentally, Tatra did initiate legal action after the Second World War, and the matter was eventually resolved in 1961 when Volkswagen was ordered to pay three million Deutschmarks in damages to the Czech car-maker.

But why did Porsche choose such a configuration for his interpretation of the people's car? It was essentially an issue of packaging: if the engine was going to be tucked under the rear seat, then the amount of space available for people and luggage for a given length of car would be considerable. When the powerpack is in the front, people have to sit aft of it, with stretched-out legs resting at the firewall (the panel separating the engine and the passenger area) behind the engine. Whereas with a rear-engine layout, the front 'firewall' could move decidedly forward, allowing for greater interior volume.

In the meantime, a factory was set up in Wolfsburg. Some 300,000 Germans enthusiastically began subscribing to the saving scheme for the people's car, but their money was soon diverted to the war effort. Porsche himself got involved in the development of the military version of his car, the famous Kübelwagen, which became the German equivalent of the Jeep.

In the aftermath of the World War, in 1945, the Wolfsburg factory, completely destroyed, was in the British occupied zone. The English Major Ivan Hirst tried to restart some activities by creating a repair shop for military vehicles. Without much hope, he assembled two Volkswagens from scattered pieces. Finally, he got the assembly line going again. As part of war reparation, the site and the car were offered to Ford, which refused, deciding that both the factory and product were worthless.

As Germany needed foreign currency after WWII, the decision was taken to export the Volkswagen. The Germans nicknamed it Käfer, the German word for cockroach. In the United States, it became the Beetle.

Initially sales were slow. Enter American advertising genius Bill Bernbach. Refusing to patronise the consumer, Bernbach put in humour and modesty. He described the Beetle as a car that was "ugly" and a "failure", and introduced the idea of low power, low consumption and minimal maintenance, and asked buyers to "think small" when the rest of American industry were outdoing each other in making increasingly bigger cars.

The Volkswagen Beetle, a survivor of World War II. © Makarand Baokar

For ten years, the 250 ads of this self-effacing brand published in the American press boosted sales of the Volkswagen Beetle to 462,000 units sold in 1968, a record year. Voted the best ad campaign of all time, Volkswagen's counter-culture ads were used everywhere. In France, in the 1970s the slogan "In an effort to be outdated, it ends up being fashionable" made the Coccinelle ('ladybird' in French) a resounding success. And by the time the Walt Disney film 'The Love Bug' was released in 1968, the Volkswagen Beetle was a worldwide phenomenon. Four other feature films and a TV series followed and by 1972 the sales record of the Ford Model T had been surpassed.

That could have been the end of its history. The Beetle was by then over thirty years old and out of date. The competition offered better performance, lower fuel consumption and greater comfort. Volkswagen was preparing its replacement, the Golf. The 'old', logically, had to go. Well, no! Instead, it became a bestseller in emerging countries, where the twin virtues of robustness and ease of maintenance worked perfectly for people suffering bad roads and enjoying limited resources. In Brazil and Mexico the Beetle found new fans. Of even greater surprise was its continuing success in Europe and the United States. Its retro aesthetic look seemed to be back in fashion, the advertising campaigns became increasingly funnier and some technical improvements, including a tad more powerful engine, contributed to delaying the car's retirement.

But thanks to the success of the Golf and concerns regarding economies of scale, the Beetle finally bid adieu to the Wolfsburg factory in January 1978. Yet the story continued in several other production outposts, ending, finally, in Mexico, at Volkswagen's Puebla plant, in 2003.

With 21,529,464 units made, the Volkswagen Beetle remains the greatest people's car ever. One mad man's dream became the preferred set of wheels for 21 million others across the planet.

But even as this mad man (who, incidentally, couldn't drive) was driving Porsche's idea of a people's car, across the Maginot Line, in France, another more advanced people's car was also moving from the drawing board to … no, not to a French driveway, but to an underground basement!

André Citroën had always dreamt of a revolutionary car for the people. The Traction Avant, which was developed just before Citroën lost control of his company, was the car designed to save Citroën from financial disaster post-Depression by making a car for the middle-class. The Citroën Traction Avant was the PV, the *petite voiture*, or 'small car', but the follow-up to that would be the TPV — *toute petite voiture* — or 'very small car'.

Sadly, Citroën didn't live to see the TPV. Instead, Pierre Michelin, who had taken over Citroën, gave the go-ahead for the TPV to Pierre-Jules Boulanger, who had been brought in as the boss of the car firm. The idea was to create a car for the rural and low-income masses, the primary objective being that of increasing Michelin's tyre sales!

Based on a survey amongst the targeted audience, Boulanger drew up very precise specifications and rather stringent objectives: a really small car with four seats, the possibility of transporting 50 kg of baggage, front wheel drive (like the Traction Avant), a max speed of 60 km/h, a three-speed gearbox, easy to maintain, run 30 kilometres to a litre of petrol, fit into the French taxable horsepower rating of 2hp, and, most importantly, a suspension system that would allow the car to cross a ploughed field with a basket of eggs, without breaking one! The car had to be easy to drive, so that a beginner or a woman could use it. Nothing ostentatious, just "four wheels under an umbrella", the slogan that was used in the late 1960s, pretty much summing up the spirit of what Boulanger expected.

André Lefebvre, who had, at very short notice, developed the Traction Avant, was given the brief and a free hand. Lefebvre brought in Pierre Meyer and Alphonse Forceau, the latter specifically to develop the rather demanding requirement of suspension design, Jean Muratet came in for the body construction, though the styling of the car was left to a bright young sculptor, Flaminio Bertoni, who had already impressed with the low and stylish Traction Avant.

The team had carte blanche as long as it was not aesthetic or performance enhancing, as these two were seen as features that would add to the costs of making the car. The project explored all possible avenues, even rare materials such as magnesium, the doors opening in a circular fashion (not exactly safe for the fingers!), mica windows to reduce weight, and even experimented with a lighting device inspired by fireflies!

As many as 47 iterations and prototypes were developed, each evaluated by Boulanger, before the definitive was selected. When a solution led to an excess of comfort, it was rejected, as by reducing the comfort, costs could be curtailed. Two ideas, however, were retained: the heater that the engineers had crafted for winter testing, which consisted of a sleeve that recovered heat from the exhaust pipe; and the door trim and upholstery that Madame Boulanger had suggested to her husband.

So, just at the dawn of World War II, by September 01, 1939, 250 cars were assembled and ready at the Citroën factory. A few days earlier, on August 28, the cars had received their homologation certificate too. Citroën's TPV was all set to roll out into the market.

But with war clouds on the horizon, the Paris motor show was cancelled. Then, with the invasion of France by the Germans, orders came from the boss, Pierre-Jules Boulanger to destroy all the cars. Some prototypes were hidden in the attic of the test centre, some in the basement of the design office.

During the occupation, the Germans, who were aware of the TPV project, asked Pierre Boulanger for the designs, in exchange offering the designs of the KdF-Wagen. But the Citroën boss refused. Boulanger made every effort to sabotage the German effort at getting the prototypes and the drawings ... and, in the meantime, rethink several aspects of the design. The original aluminium body was abandoned in favour of steel. The 375cc water-cooled engine was redesigned into an air-cooled unit by Walter Becchia, who also managed to incorporate a four-speed gearbox within the housing of the three-speeder. But the most innovative aspect was the suspension design which allowed for amazing travel, comfort, and a hilarious amount of body roll in cornering even while ensuring that the car wouldn't tip over easily.

Finally, on October 07, 1948, at the Paris motor show, the Citroën 2CV (2hp in French) — as it was eventually called — was unveiled. Show-goers to the Paris show took to the strange little car, even if, like the Ford Model T, you could order your Citroën 2CV in any colour as long as it was grey!

But the press disliked it. An American motoring journalist wondered whether the car came with a can opener. A correspondent for British car magazine *Autocar* wrote that the 2CV was the "work of a designer who has kissed the lash of austerity with almost masochistic fervour."

With material supply a problem, the start for the Citroën 2CV was rather slow. In its first full year of production, in 1950, just 6,196 cars were produced. In 1951, production went up to 16,288, and a van version was introduced. In 1952, Citroën managed to shift a little over 21,000 2CVs, plus over 7,000 of the van, the Fourgonette. As the numbers kept going up, so did the number of markets and then, model derivatives. By the time the 2CVs production came to an end in 1990, 42 years after its launch, over 3.8 million of 'the umbrella on wheels' had been produced, not counting the

1.2 million-plus vans, plus the derivatives, the Ami (1.8 million), the Dyane (1.4 million), the Acadiane (253,392) and the Mehari (144,953). A total of 8.7 million-plus!

The Citroën 2CV – an umbrella on wheels. © Marianne Debiesse

In time came grudging admiration from the press. Eminent journalist and author L J K Setright described the 2CV as, "the most intelligent application of minimalism ever to succeed as a car."

And minimalism was the order of the day in Europe's post-war situation where the devastation of a horrific war had decimated the automobile population drastically. For instance, of the two million-plus car parc [cars on the road] in France before World War II, barely a hundred thousand-odd had survived the war. Though there was a desperate demand for personal transportation, there was also the situation of severely diminished purchasing power.

So, into the immediate post-war scenario, Europe had two people's cars with distinctively different approaches: the Citroën 2CV with its front-engine, front-wheel-drive arrangement, and the Volkswagn Beetle with its rear-engine, rear-wheel-drive configuration. From these two approaches followed a whole slew of small people's cars to meet the post-war demand for cheap, and not necessarily cheerful, personal transportation.

In Germany, following Volkswagen's lead, BMW, Glas, NSU and several others came up with small, inexpensive, rear-engine cars that aimed to be cars for the masses. The BMW 600 was a BMW-developed derivative of the BMW-Isetta, which BMW was making — to jumpstart car manufacturing — under licence from Italian scooter-maker, Isetta. Yes, BMW — twice over — had very humble beginnings!

Powered by a small 586cc 20bhp two-cylinder engine, the BMW 600 was most unusual in having a door at the front for access to the front seats; access to the rear bench was through a door on the right side of the car. Glas, NSU and others also developed intriguing and strange features and, like the BMW-Isetta and the BMW 600, were referred to as 'bubblecars'.

But the most successful champions of the rear-engine cause were from two other European countries: the Renault 4CV from France and the Fiat 500 from Italy. Renault went on to make over a million of the 4CV, a small four-door car produced from 1947 to 1961 and Fiat made over five million of the original 500 (1957-1975).

The Fiat 500 – Italy's people's car.
© Makarand Baokar

But Renault also realised, when it saw arch rival Citroën's 2CV, that the way of the future was the front-engine, front-wheel-drive way, so it developed the 4L as a replacement of the 4CV, which, in many ways, was more than a worthy competitor to the Citroën 2CV, selling over eight million between 1961, when it was launched, till end of production in 1992.

But the car that made the best case for the front-engine front-wheel drive cause was the Mini. Taking the Citroën 2CV idea one step forward, Turkish-born Greek-British designer Alec Issigonis came up with a car that was very compact (some 73cm shorter than the 2CV), with a decent-sized engine (a 38bhp 848cc transverse four-cylinder) driving the front wheels, with enough space for four average-sized adults in a comfortable fit, in a car that was a whisker over three metres long. Initially badged both a Morris Mini-Minor and Austin Seven (as the rival British car-makers had merged by 1952), the car was eventually sold as just the Mini.

The Mini – Britain's most important post-war people's car. © Makarand Baokar

Into the 1960s, it became obvious the Mini template was the way to go: the Mini was small, yet had amazing space for its size and was a delight to drive; with handling that put many bigger, sportier cars to shame; winning rallies and races yet managing to be chic in a very 'classless' sort of way. Over the next two decades, most small cars became front-engine front-wheel-drivers. The Mini, in its 41-year-long run —1959 to 2000 — sold over 5.3 million, making it the most successful British car ever.

In retrospect, though, most agree that the Citroën 2CV was the most original people's car in the 20th century. The Ford Model T was not minimalist and was a regular mid-sized car for its day. The product was not innovative, the production process was. The Volkswagen Beetle wasn't original either — there is evidence that Porsche was more than inspired by the Tatra T77 and T97. The Beetle's success was in being so different from the cars selling in the world's biggest market then, the USA. Plus, the amusing, tongue-in-cheek, self-deprecating advertising that went with the growing counter-culture of the 1960s. In that sense, the 2CV too became a darling of the sub-culture of the 1960s and 1970s, but more so because of its in-your-face minimalism than any other reason. For somewhat similar reasons, the Mini and the Fiat 500 too became cult favourites in their respective nations, populist transport for the anti-bourgeois bohemian set, automobiles that were affordable wheels and a ticket to personal freedom during the swinging 1960s and the psychedelic 1970s.

But all these cars had one thing in common: they all became *the* engines of growth for the very important car industries of the USA, France, Britain, Germany and Italy. These people's cars not only gave the people of these countries their very first set of wheels, they acted as drivers for the hugely important automotive industries of these nations and, in the process, convinced leaders of many developing nations, whether democratic or authoritarian, that that's what was needed for economic growth, that every country must have, at the least, a 'people's car'.

The same idea came to fruition even in India, but before that the country needed to get independence, get a life, and get an automobile industry.

Chapter II
Motoring Pioneers

It is quite possible that the first Indian to run his own car may have been Kumar Shri Ranjitsinhji Vibhaji Jadeja, otherwise and better known as Ranji. Yes, the same Ranji of Ranji Trophy fame. Ranji became Maharaja of Nawanagar, a minor princely state in Gujarat, in 1907. But he was already famous as one of the world's finest cricketers. Earlier, when he was in the UK studying at Cambridge, between the years 1889 and 1893, he acquired a very early automobile that was the cause of much excitement and consternation in that quaint university town.

The first car to be acquired by an Indian in India, however, was the steam-powered, two-cylinder three-wheeler, French De Dion-Bouton that Maharaja Rajinder Singh of Patiala purchased in 1892. In 1893, an Olds steam car, made by the pioneering American, Ransom E Olds, was shipped to India for an unknown customer. The car never arrived on Indian shores as the ship carrying it sank. However, this car is still recorded as the first ever export of an automobile from the USA.

In 1897, a Benz was imported by a British gentleman in India, J B Foster of the company Greaves Cotton & Co. The car arrived in the city of Bombay. Soon, cars were brought into Calcutta too, the capital of British India then. In 1898, a European firm imported three 'horseless carriages' that found their way to some petty royal states and the Indian maharajas' love affair with the automobile began.

But that love affair was limited mainly to the maharajas, the wealthier English families residing in India, and the merchant class from the cities of Calcutta and Bombay, as the motor car then was not only a very pricey piece of kit, but also way beyond what most of India's 200 million poor could afford. With an average annual per capita income of less than Rs 200 and a car costing upward of Rs 5,000 (the equivalent of 25 years salary) for most of India's 238 million people at the turn of the century, the automobile was nothing more than an object of reverential curiosity. For the very rich, it was nothing more than a plaything.

Yet, car-makers thronged to India, appointing dealerships in the port cities of Bombay, Calcutta and Madras. All kinds of cars were imported — electric, steam and petrol — as it still was not clear which would work best. The European manufacturers took the lead, specifically the French, with importers and dealers selling Chenard et Walcker, Clément-Bayard, Darracq, De Dion-Bouton, Delaunay-Belleville, Hotchkiss, Mors, Panhard et Levassor, Renault and Richard-Brasier (later Brasier). Soon the fledgling British industry laid claim to the important market represented by the crown jewel of the Empire, with marques like Daimler (made under licence from Germany in the UK), Humber, Lanchester, Napier, Rolls-Royce, Standard, Star, Thornycroft, Vulcan, Wolseley and Wolseley-Siddeley. American brands like Olds, Rambler, Cadillac and Studebaker followed, as did Italian cars like the Fiat and Bianchi, and the Minerva from Belgium.

This Darracq 1905 from France was an early import into India. © Gautam Sen

In most cases, complete (fully assembled and ready for the road) cars were imported and sold, though many of the wealthier princes preferred to 'shop' for their cars during their annual visits and summer holidays to England, whence a fleet of cars would be selected and purchased, to be shipped to their principalities back home in India.

Several car-makers also shipped out their cars in chassis-mechanical form, to be 'bodied' in India. In fact, the coach-building of bodies was perhaps the earliest automotive manufacturing activity in India. With a tradition of skilled woodwork, the art of coach-building had an early history in India.

The oldest known was that of Steuart & Co, from Calcutta. Established in 1775, Steuart & Co built up a very fine reputation for their ornate and elaborate state coaches, elephant howdahs and a variety of horse carriages — many sumptuously decorated with gold, silver and ivory. On special appointment to the Prince of Wales and the Viceroy of India, the list of Steuart & Co's clientele included the who's who of royal India.

An ad for Steuart & Co. Courtesy Calcutta Motor Sport Club, CMSC

Motoring Pioneers

With the coming of the horseless carriage, Steuart & Co realized it had no choice but to embrace the new-fangled device, thus moving into the world of coach-building of automobiles — a service it started offering at the advent of the motoring era in India. It also saw an opportunity in distributing cars and took up the distribution of several European brands including the British Thornycroft. Established in 1862, by John I Thornycroft, with plans to make steam-driven coaches, Thornycroft made its first steam car in 1895. By 1902, the company had moved to internal combustion power, launching its first car in 1903. Starting with a small 10hp two-cylinder car, Thornycroft moved to making bigger cars — a 20hp four-cylinder, followed by a magnificent 6975cc six-cylinder car that, in later 7819cc form, was more than a worthy rival to the Rolls-Royce Silver Ghost.

Thornycroft was one of the early car-makers to offer complete cars with standard bodies already installed. Naturally, complete cars cost more than the bare chassis and the price of a complete car from Thornycroft depended on the bodywork specified, of which there were several options, including the Victoria, the landaulette, limousine, Roi-des-Belges, *et al.* Thornycroft also sold bare chassis to those who wanted them. So Steuart & Co, other than selling complete Thornycrofts, imported significant numbers in chassis-mechanical form, building at its facilities in Calcutta, the car bodies as per customer requirements or following the styles established by the car-maker.

As the popularity of the motor car increased, Steuart & Co soon had competition. Not very far from its facilities on Calcutta's chic Bentinck Street, French Motor Car Company Limited established a showroom, followed by a coach-building facility a few kilometres away. Starting with French car brands such as Panhard et Levassor, Mors and Berliet, French Motor soon added other European and American brands, including Armstrong-Siddeley, Bean, Bianchi, Minerva, Cadillac and Studebaker. Other than selling complete cars, French Motor also began bodying chassis-mechanicals shipped out by many of the car-makers. Business was so good that French Motor opened an important branch in Bombay soon thereafter. The combined staff of the two facilities added up to over 600 people.

In Bombay though, French Motor Car Company had another worthy rival in Fort Coach Factory. Established in 1877, the Fort Coach Factory was founded by a Parsi gentleman by the name of Pestonji B Press. With a background in art from Bombay's famous Sir Jamsetjee Jeejeebhoy (JJ) School of Art, Press had an eye for aesthetics and soon built up a fine reputation for making exquisitely crafted horse carriages. With the advent of the motor car, Fort Coach Factory moved to specialise in car-body-building too, starting with bodies for the French Brasier cars, imported in chassis-mechanical form.

But the coach-builder who was arguably the most successful in India, was Simpson & Co from Madras. Established in 1840, Simpson & Co eventually was on special appointment to none other than King Edward VII and the Nizam of Hyderabad, having made fabulous carriages, both horse and elephant-drawn. In the motoring era, other than distributing and selling cars (starting with steam-driven ones), Simpson & Co built hundreds of coach-built cars based on chassis from 'middle-market' car-makers such as Darracq to the prestigious Daimler, Delaunay-Belleville, Napier and Rolls-Royce.

Yet Simpson never got into full-fledged manufacturing of cars, let alone assembling, unlike several other coach-builders-turned-carmakers such as Holden in Australia and Vanden Plas in the UK. Holden, founded by James Alexander Holden in 1856, as a saddlery, became a coach-builder by 1914 and eventually a full-fledged assembler of GM cars. In 1931, Holden became a part of the GM empire. But the brand remains Australia's most important. Vanden Plas' history goes back to the horse-drawn carriage era, then to coach-building of cars, to series manufacturing of bodies, eventually becoming a brand within the British Motor Corporation from 1957.

Instead, Simpson & Co took the route to becoming a manufacturer of diesel engines and trucks after Indian independence. Today, it specialises in manufacturing and supplying a family of diesel engines for truck, tractor, forklift, marine and genset (generator) applications. Sadly, Steuart & Co, French Motor Car Company and Fort Coach Factory did not survive the move away from coach-built cars to series production automobiles as the major car-makers set up their own assembly lines with local partners.

But before that there were efforts to make an indigenous car. Samuel John Green, a director at Simpson & Co, soon after joining the company in 1902, attempted to make a car with an engine, pump, boiler and gears, all made in-house at Simpson's facilities. But in the true sense of the term, the

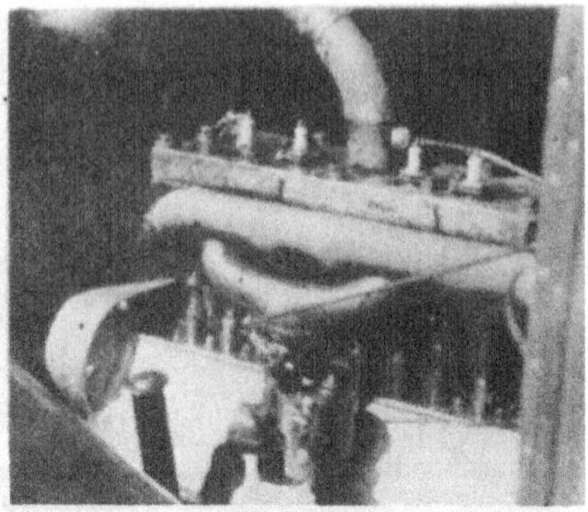

১. বিপিনবিহারী দাস ও তাঁর 'স্বদেশী' গাড়ি ২. 'স্বদেশী'র ইঞ্জিন ৩. নিজের কারখানায় বিপিনবিহারী ৪. 'স্বদেশী'র স্টিয়ারিং ইত্যাদি ৫. বালিগঞ্জ ফাঁড়ির কাছে 'স্বদেশী' ও বিপিনবিহারী

Clock-wise from top *Bipin Behari Das and his 'Swadeshi' car; the engine of the Swadeshi; the Swadeshi's steering wheel; Bipin Behari and the Swadeshi at Ballygunge; and Bipin Behari at his workshop.* Courtesy Calcutta Motor Sport Club, CMSC credit for building the first car goes to A A Bachmann from Calcutta (now Kolkata). An early enthusiast who had participated in the first automobile meet at Calcutta in 1904, Bachmann was a trader in cars and a joint patentee of an ingenious friction-and-belt differential gear for cars (granted in 1910). In 1912, he started manufacturing cars, though the specifications of the Bachmanns are not known.

A wholly indigenous automobile is claimed to have been made by a Calcutta-based entrepreneur, Bipin Behari Das, in the 1920s. Named *Swadeshi*, echoing Mahatma Gandhi's self-reliance sentiments, this car was made from locally available materials and lashed together in a garage with very basic equipment. Whether it was grounds-up all indigenous or whether it was based on existing mechanicals is not known. What is known is that the car was approved by the police and the registration number '35977' was allotted to it. The car was eventually purchased by the Calcutta Corporation for Rs 3,000. Calcutta Corporation was known to have financially supported Das, who built a car for Benares Hindu University, which was used by Motilal Nehru and Pundit Madan Mohan Malviya. Nothing more seems to have come of the project though.

'Made in India'

Thus it came to pass, some years later, that the first set of 'made in India' cars were American automobiles. Though Ford Motor Company was first off the block in incorporating Ford India on 31 July 1926, the company took a while to set up its first assembly plant in Bombay, in the suburb of Bhandup, in 1931, followed a few months later by similar facilities in the port cities of Calcutta and Madras.

In 1926, the first year of Ford's direct operations in India, 4,152 Model Ts were sold over the calendar year, accounting for a third of all Fords sold in South and South-East Asia that year, and representing an almost 13 per cent improvement over the preceding year, when Ford was selling through distributors. Of course, all of these were direct imports. By 1930, the number of Fords sold in India had gone up to 5,107 (all imported), by 1935 to 5,444 (by now, mostly assembled), and then to 7,011 by 1940. In fact, in the WWII years, 1940 to 1946, Ford assembled and delivered as many as 114,485 defence vehicles to the Allied forces in India from its three assembly plants.

Ford Canada had the distribution rights for the British colonies and had set up assembly facilities in Africa as early as 1923 and in Australia in 1925. But before Ford got going, in 1928, two assembly plants were up and rolling in India, both by rival American car-makers: General Motors and Graham-Paige. Founded as recently as 1927 by the brothers Joseph and Robert Graham and their Canadian partner Ray Austin, by taking over Paige-Detroit Motor Company, Graham-Paige rapidly went international. Of the several assembly facilities that Graham-Paige set up across the globe, one was in Calcutta, when wealthy Barrister Sushil Chandra Chaudhury decided to invest in a car assembly facility.

Chaudhury's nephew, Anil Kumar Mitra, had studied automobile engineering in the USA and worked with Ford before signing up with fledgling car-maker Graham-Paige and heading back home to Calcutta to set up the assembly facilities at 2 Rowland Road, where the Chaudhurys already had a tyre service centre — International Tyre & Motor Company. Initially, Graham-Paige was very successful, selling well as taxis because of the car's rather compact turning circle, but the investment was considerable and, within two years, International Tyre & Motor was bankrupted. Also, the supply of kits may have been affected by the Great Depression.

General Motors, on the contrary, had a successful run in India. Though its assembly facilities at Sewri, neighbouring the Bombay Port, was inaugurated in December 1928, series assembly began in earnest in 1929, with several of the many GM brands, including Buick, Chevrolet, Oakland, Oldsmobile and Pontiac, rolling off the lines at the same time. Trucks with the GMC branding were also introduced. So, by the time GM overtook Ford in 1931, the Indian market too, was contributing to the General's global lead. This was reflected in the increasing popularity of Chevrolets and Buicks in the sub-continent.

Unfortunately, numbers are unavailable for car sales in pre-war India, but somewhere out there in the statistical wilderness, there is this little figure: about 17,000 cars may have been sold in the Indian sub-continent in the year 1930. A plausible figure, given that in 1950 — the first officially available statistic — car sales in a divided and significantly impoverished India totaled 14,688 units.

An Indian-assembled Chevrolet from the 1930s, probably modified from a saloon to a tourer. © Makarand Baokar

The number of 17,000 looks impressive when compared to another piece of data, that of the total car parc of the city of Bombay in 1910 — all of 1,025 cars registered till then. Calcutta, being the capital of British India and a commercially more important city at that point of time, may have had a few hundred cars more, but it would be safe to estimate the total car parc across India could not have been more than 10,000. So, a yearly sale of 17,000 cars — almost twice the total car parc in the first decade of the automobile in India — was a not-so-insignificant number in the automotive world then. Thus, it is surprising that only one other car assembly set-up came up in the 1930s — Addison in Madras — to assemble British Wolseleys. It was a logical move as Addison was already the dealer for the Wolseleys, Morris and Riley, amongst others.

Incidentally, most British car-makers did not need to set up assembly lines as the market was, in a sense, a captive one for them. Plus, it was not in their nature to do so. Austin preferred licensing production elsewhere and Morris Motors did not have a Ford-style assembly line till 1934! For the Americans, shipping complete cars all the way from USA was expensive. Also, Ford was a pioneer in transplanting the assembly line and in international thinking with the Model T, opening plants in Canada and the UK as long ago as 1911, when Model Ts were assembled from knocked-down kits. GM, sure enough, learnt quickly.

The failure of the first Indian car assembling venture, that of International Tyre & Motor to make the Graham-Paige cars, may have been a deterrent to many Indian entrepreneurs looking to invest in the car-making business in the 1930s. Also, the Great Depression, by wiping out two-thirds of the car-makers across the globe, played a role.

Yet, it is interesting to note that in 1930, the Japanese fledgling car industry made a modest start with the manufacture of 500-odd units.

One theory on why car-assembling ventures failed in India is that Indian business houses just did not have the confidence to take on such capital-intensive projects with British 'masters' around; that it was the changes in the political environment that finally got the Indian businessmen to seriously start thinking of automobile production. But by 1940, the writing was on the wall — the beginning of the end of British imperialism and the dawn of a new order and a new era in Indian industry.

Made by Indians

Business houses like the Birlas and Walchands, both close to the Indian National Congress party, were the first movers to perceive business opportunities in automobiles. Walchand Hirachand Doshi, born in a Gujarati Jain family based in the small town of Sholapur in Maharashtra, was an enterprising young matriculate who cut his teeth in the railway construction business, but saw opportunities in shipping and then the construction of aircrafts. In fact, Hindustan Aircraft in Bangalore (later Hindustan Aeronautics Limited, HAL), was born from a chance meeting between an American aircraft executive and Walchand Hirachand on a flight to the US in 1939. Walchand's visit had to do with a possible collaboration with one of the American Big Three for the manufacture of cars. His first stop was General Motors, the biggest of the Big Three, but then GM already had its assembly set-up in Bombay, which was on full throttle. An audience with Henry Ford went well, but Ford asked him to speak to the Ford Motors' Canadian division, which was handling the British Commonwealth markets, including India. Also, Ford by then had three successful assembly operations going on in India. So Walchand Hirachand then knocked at the doors of Chrysler Corporation, a company that, in comparison to the other two giants, was almost an upstart.

Brought in to manage an ailing American car-maker called Maxwell Motor, Walter Percy Chrysler launched the first Chrysler-badged car, the Chrysler Six, in 1924. Featuring hydraulic brakes on all four wheels and a powerful six-cylinder engine, the Chrysler Six was an instant sales success. The success of the Six helped Walter buy the remaining assets of Maxwell Motor Corporation to create the Chrysler Corporation. The development of this new company was rapid. By early 1926, Chrysler had nearly 3,800 dealers across the US and a profit of US$17 million — big money then.

The success of the company continued into the following decade, with the launch of many new models, including the Imperial 80 in 1927. Not only was this car the first truly convertible Chrysler, but it was also the first model of the brand to be advertised in colour. In 1928, Chrysler founded the Plymouth brand for the entry level segment, the DeSoto for the mid-range, and bought the Dodge Brothers Company. This strategy was designed to offer a full range of vehicles along the lines of what General Motors had built up over the years since its inception in 1908.

It worked so well that by 1936, Chrysler was in second place behind GM on the US sales charts and ahead of erstwhile leader Ford Motors! Ford, with its single model-to-suit-all-needs strategy, had slipped to third place in the US marketplace, with a market share of 22 per cent. Chrysler had 25 per cent of the market, GM 43 per cent. Between them, the American Big Three had 90 per cent of the world's biggest market and more than half the world market.

The US ruled the car business worldwide. In 1938, the US produced more than 2.5 million of the world total of four million cars produced that year. Great Britain, which had become the number two car-maker in the world, overtaking France in 1929 for the first time, made 445,000 cars that year, accounting for nine per cent of world production. In the meantime, the Japanese industry had come a long way from its modest 1930 start to a total production of 24,000 units in 1938.

Since Walchand Hirachand was not taking the Japanese route of developing an auto industry with products developed in-house, it was better to tie-up with an international giant. So Chrysler Corporation and Walchand Hirachand signed up in 1940.

But then WWII intervened and car production was put aside for the demands of the war effort. Chrysler Corporation's contribution included the manufacturing of the famous Sherman tanks and Dodge made the engines for the Boeing B-29 Superfortress and military trucks. Walchand Hirachand, in the meantime, incorporated his car company, Premier Automobiles Limited, in 1941, and then had to wait patiently for the war to come to an end and car production in the US to recommence, in 1946.

It was a somewhat similar story with the Birlas. Patriarch Ghanshyam Das Birla was close to Mahatma Gandhi and a great supporter of the Quit India movement. He realised there was considerable potential in setting up an automotive assembly plant near a port. Hindustan Motors was incorporated in 1942 and established at the tiny Gujarat port town of Okha. The war and the consequent diversion of efforts into war materials had all the car majors of Europe busy. So Birla and Hindustan Motors had to wait.

With the possibility of India's independence around the corner and an altogether new regime in power, Birla and Walchand Hirachand decided to wait things out. Other industrial groups also saw opportunities and automotive enthusiast, J R D Tata, decided to get the Tata Group to incorporate Tata Engineering & Locomotive Company Limited (TELCO) in 1945, with plans to make locomotives and vehicles. In Ludhiana, a company called Mahindra & Mohammed was incorporated in 1945 to trade in steel. With India's independence from British rule came the partition of the country into India and Pakistan. It caused the greatest transmigration in the history of humankind, with more than 12 million people crossing the borders between India and West Pakistan and East Pakistan.

It was only after things stabilised that the business of car-making and assembly was once again considered. By 1948, Mahindra & Mohammed had become Mahindra & Mahindra, as business partner Ghulam Mohammed left for Pakistan to become Pakistan's first finance minister. Hindustan Motors migrated to West Bengal in 1948 and set up a factory in a town called Uttarpara, 15 kilometres north of Calcutta.

Motoring Pioneers

A Chevrolet Fleetmaster from 1947, assembled at GM's Indian facilities. This car is currently owned by Robert Niedu, an Italian entrepreneur living in Jodhpur with his family, at the haveli in the background. © Makarand Baokar

Premier Automobiles were first off the block, getting its factory in the Bombay suburb of Kurla up and running by 1949, when the first set of Dodge, Plymouth and DeSoto-badged cars rolled out, along with the first batch of trucks with Dodge, Fargo and DeSoto branding (Fargo being the brand for trucks sold by the Plymouth dealers). The same year, the assembly lines began rolling at Hindustan Motors as the company cleverly signed two deals — one with American car-maker Studebaker and the other with British manufacturer Morris Motors. One of the older American car-makers, Studebaker had a storied history. Starting with horse carriages in 1852, Studebaker migrated to car-making in 1902, initially making electric cars before moving to petrol-engine automobiles. During WWII, even though Studebaker was deeply implicated in the war effort, making amphibious vehicles, many of which found their way to India, they were also preparing for the anticipated post-war market. Thus, in 1946, Studebaker was ready with a brand new range of cars, which it launched with the famous slogan, 'First by far with a post-war car'. This may have attracted Hindustan Motors to ink an assembly deal with Studebaker.

A Buick Eight from 1948, also assembled by GM in India. © Makarand Baokar

But the Studebakers, in typical American fashion, were big six-cylinder-engine cars, expensive to buy and even more expensive to run. So a smaller car that would appeal to the middle class was what Hindustan Motors needed. As a complement to the Studebakers, Hindustan Motors decided to tie-up with British car-maker Morris Motors, who had been selling its inexpensive 'middle-class' models in India successfully.

A Studebaker from late 1956, assembled by HM at its Uttarpara facilities. © Makarand Baokar

Headquartered in Oxford, William Morris created his first car in 1910 — a two-seater buggy. Later, he made cars with American engines and gearboxes and at the end of WW I, sourced its mechanicals from French car-maker Hotchkiss. Producing reliable cars at bargain prices, Morris became Britain's number one car-maker by 1925. In 1928, Morris launched the Minor, a worthy rival to the Austin 7. The Minor sold well in India too.

Morris acquired the MG sports car brand, Wolseley and Riley, in the 1930s. After WWII, Morris re-introduced the Ten, a small car it had first launched in 1939. Hindustan Motors began its Morris assembly operations with the Ten, launched in India as the Hindustan 10, in 1949.

The Hindustan 10, one of the first cars with an Indian 'badge'. © Makarand Baokar

Shortly thereafter, in 1950, Hindustan replaced the 10 with the 'Baby Hindustan' (the recently unveiled Morris Minor), and complemented it with the slightly bigger Hindustan Fouteen (the Morris Oxford). The Minor retailed for Rs 8,025 then and the Hindustan Fourteen went for Rs 10,085.

Hindustan Motors also imported and sold the Morris Six — an Oxford with a longer bonnet, incorporating a bigger six-cylinder engine — at a pricier Rs 13,325. In comparison, the Studebaker Champions, the Deluxe Sedan and the Regal Deluxe Sedan, both assembled by HM, retailed for a relatively reasonable Rs 11,910 and Rs 12,210. In 1954, the Hindustan Fourteen was replaced by the Hindustan Landmaster from the Morris Oxford Series II. The Landmaster remained in production till 1957, when it was replaced by the Ambassador of the Morris Oxford Series III, launched in 1956 in the UK, but not in India till 1957. In 1959, when the Oxford Series III was replaced by the all-new Farina-designed Series V, the tooling of the Series III was shipped to India to become the Ambassador. Since then the Ambassador has been made…and made…and at the time of writing, is still being made…

A Hindustan Landmaster, beautifully restored by current owner Karl Bhote.
© Karl Bhote

Though Premier Automobiles started off before Hindustan Motors, its product planning was not as good. Till 1951, PAL was assembling the Chrysler Corporation cars, the DeSoto Diplomat, the Dodge Kingsway and the Plymouth DeLuxe — all retailing for similar prices: Rs 15,275, Rs 15,175 and Rs 15,075 respectively — thus cutting into each other's sales in their rather narrow market segment.

Dodge Car.

Fiat Car.

Premier's two early offerings, the Dodge and Fiat 500C. Courtesy Premier Ltd

A Dodge Kingsway from 1957, assembled by Premier Automobiles.
© Makarand Baokar

True, Premier had a lucrative truck-assembling activity, making Dodge, Fargo and DeSotos, but more important was the business of selling Fiat cars in India. Though Walchand Hirachand's first meeting with Chrysler Corporation is well documented, how the Indian car assembler got to tie up with Fiat of Italy is not well known. What is known is that Premier had also approached Volkswagen, among others. If inexpensive, smaller cars had to be made, then the obvious collaborators were the European car-makers. Amongst them, Fiat already had the reputation of making some of the most popular smaller people's cars, specifically the legendary 500 Topolino. Since 1936, it was one of the smallest and cheapest cars made in Europe.

Initially, completely built-up Fiat 500C Topolinos and 1100s were imported and sold for Rs 5,400 and Rs 9,725 respectively, with the former having the enviable reputation of being India's cheapest car then and closest to a people's car. But it had two serious shortcomings — the Fiat 500C had just two doors and the four seats fit only two adults and two children, no more — features unacceptable to chauffeur-driven, joint-family Indians.

The Fiat 500C on Premier's assembly line. Courtesy Premier Ltd

Assembling and selling a bigger four-door car, like the HM models, was a more sensible alternative, so Premier inked a fresh deal with Fiat. At the outset, PAL assembled the two-door 500C, starting 1951, but soon followed that up by introducing the all-new 1100 in 1954, a year after it had been launched in Italy. Till 1965, as Fiat in Italy kept updating the 1100 with cosmetic upgrades, the Premier-assembled Fiats in India too, kept getting facelifts. In 1965, Premier introduced the Fiat 1100 D or Delight. This was the Italian Fiat 1100D from 1962, which was made in Italy till 1966. The tooling was shipped to India soon thereafter, and like the Ambassador, the Fiat 1100D remained in production from 1966… to 2000!

The Fiat 500C, India's cheapest car then. © Kunal Khadse/Adil Jal Darukhanawala

In the meantime, TELCO went into the business of manufacturing trucks in a factory in Jamshedpur, some 300 kilometres west of Calcutta, after it had got into a technical collaboration with Daimler-Benz, Germany, in 1954. A prominent importer in Calcutta, Dewar's Garage, also waded into the assembly business, bolting together and selling British brands such as Singer and Rover. Mahindra & Mahindra inked a deal with Willys to assemble the Willys Jeep CJ3B, under licence in India, in a factory located in north Bombay; this was sold as the brand 'Mahindra Jeep'. Even a few big American Kaisers were being assembled at a facility in Bhandup, Bombay.

A Fiat 1100 Super Select from 1962. © Makarand Baokar

The Taxis of Mumbai

Even more than a decade after the Fiat 1100-derived Premier ceased production, the streets of India's biggest city, Mumbai (Bombay), continued to be ruled by thousands of black-and-yellow Padminis. These inexpensive run-abouts have been very popular with the taxi trade. But a recent ruling by the local government disallowed taxi services using cars that were 20 or more years old or more. This resulted in the phased scrapping of the Premier Padminis; 2019 will spell the end of the Fiat taxis and signal the end of an era....

Madras too (now Chennai), had its fair share of automotive assembling action. Raghunandan Saran, a Madras-based industrialist and friend of India's first Prime Minister, Jawaharlal Nehru, was advised by the latter to look at car-making. Saran set up a plant to assemble Austin cars, naming the company Ashok Motors. Starting operations in 1948, Ashok Motors assembled and sold the Austin A40 from 1949. The Austin A30 was also considered but apart from receiving a prototype in 1953, no progress was made on this, as Leyland trucks became the focus.

For the Indian automobile industry, 1951 was a landmark year, when the Industries (Development and Regulation) Act was instituted. The Act required entrepreneurs to get a licence to set up a new unit, expand it, or change the product mix. The purpose of this system of licensing was to create a planned pattern of investment, minimise resource waste, counteract monopolies and the concentration of wealth, and maintain regional balances. All very good intentions, but then good intentions are one thing and execution (specifically in India), is another.

In the automotive sector, execution meant the right to make different kinds of automobiles was restricted to a handful of manufacturers. Mahindra & Mahindra were told to concentrate on Jeeps. TELCO was given a licence to make only trucks. And so was Ashok Motors, which had to give up its assembling of Austins. Austin, in the meantime, merged with HM's collaborator, Morris and the Nuffied Organisation, in 1952, to form the British Motor Corporation (BMC). Similarly, Bombay-based Automotive Products of India (API), which had an assembly operation for another British brand, Hillman, from the Rootes group, from 1949, were given a licence to manufacture scooters. So, in 1955, API switched to the manufacturing of Lambretta scooters (instead of assembling the Hillman), in collaboration with Innocenti, Italy.

If America had its Big Three, India had its Little Three — Hindustan Motors, Premier Automobiles and a Madras-based assembler, Union Motor Company, which set up an assembly line in 1949, after getting a licence to make cars. The government's thinking was that there should be one car-maker at most in each of India's three major ports.

Union Motors had tied-up with British car-maker Standard Motor Company. Another pioneering British brand, Standard, was founded in 1903, in Coventry, by Reginald W Maudslay. Standard made an early impression on the royal families of India and secured the order to supply cars for the Delhi Durbar of 1911, when King George V and Queen Mary travelled to Delhi for their coronation as Emperor and Empress of India. As many as 60 Standard cars and ten trucks were shipped out to India.

From 1929, Standard was also supplying chassis-mechanicals to the Swallow Sidecar Company, which later became Jaguar. After WWII, Standard acquired Triumph Motor Company in 1944 and it was decided that the

Triumph moniker would be used for the sportier cars and Standard for the regular saloons. The first post-war car Standard launched was the Vanguard, in 1948. Union Motors began its assembly operations with the Vanguard saloon in 1948, and then went on to introduce the smaller saloon, the Standard 8, followed by the 10 and, subsequently the Pennant. Though the Vanguard was not exactly cheap at Rs 10,975, the 8 retailed at a relatively reasonable Rs 6,500 in 1950.

At some point the name of the company changed from Union Motors to Standard Motors Product of India Limited, though Standard UK had no stake in the Indian company. Not unlike Hindustan Motors and Premier Automobiles, Standard Motors too, narrowed operations to become a one product manufacturer by 1962, when it launched the Triumph Herald, badged as Standard Herald. Though the Standard badge was last used in 1963 in the UK, it carried on in India till 1987, with the Herald and its progenitor, the Gazel, being made till 1977.

The Standard Herald – this one is the very rare, four-door version specific to India.
© Makarand Baokar

Why did this product stagnation happen? The simple answer: government policies. With the Industrial Act of 1951 and the beginnings of what is now referred to as Licence Raj, just three manufacturers in India — Hindustan, Premier and Standard — had licences to make cars. Then, in 1954, the government realised there were 60-odd models of cars selling in a market that barely totaled 20,000 sales per year. Clearly, with competition from imported vehicles, the fledgling auto industry could not survive. Also the import of automobiles and components was a serious drain on foreign exchange. With self-sufficiency as the new mantra of Nehru's government, it was decided in 1954, that high import tariffs for automobiles and components was necessary to support localisation and the growth of the indigenous automobile industry. Consequently, both GM and Ford stopped their assembly operations in India that year.

With the eventual objective of complete localisation, the three Indian car-makers decided that it would be prudent to concentrate on one specific model, for which they acquired the tooling from their respective collaborators. It also made sense to phase out the assembling of the slow-selling Studebakers and Chrysler Corporation cars, which is what Hindustan Motors and Premier Automobiles did, respectively, by the end of the 1950s.

From the beginning of the 1960s, it was a case of 'you can have any car you please as long as it is a Hindustan Ambassador or Fiat 1100 or Standard Herald'. Most people opted for the first two as the Herald came in a restrictive two-door form till the introduction of the Herald Mark III in 1968, a four-door model unique to India. Yet the Ambassador and Fiat 1100 remained the dominant models, accounting for almost 90 per cent of the Indian car market in the 1960s and 1970s — an oligopolistic situation symptomatic of crony-capitalism that led to industrial stagnation and constrained economic growth for the country.

Chapter III
In Desultory Pursuit of an Indian People's Car

Less than a decade after the Industrial (Development and Regulation) Act of 1951 was enacted, the year when 'Licence Raj' came into being, many of the Who's Who in the corridors of power began to question whether the system actually benefitted the people. One school of thought was that passenger cars per se should be a low priority item for industrial growth and progress; that two-wheelers, buses and commercial vehicles were significantly more important than cars, a product available only to the rich then. At the same time, politicians expressed concern about the cost and prices of the cars being made. At Rs 12,000 for the Hindustan Ambassador (Rs 11,554 plus taxes), the car then represented 40 years salary for an average Indian!

Though the Fiat 1100 Select was marginally cheaper at Rs 10,566 plus taxes, and the Standard Ten retailed at around Rs 10,000 (Rs 9,988 ex-factory), they were all expensive luxuries beyond the reach of the Indian masses. Though localisation had just about begun, it was still a long way from being a situation where the import of components was no longer a drain on foreign exchange which had become scarce by the late 1950s. In 1959, with the country facing a serious foreign exchange crunch, the Nehru government looked at ways to cut imports., *Swadeshi*, other than being a populist ideological buzzword, was a necessity, so the government set up a committee in April 1959, under L K Jha, an additional secretary in the commerce ministry, to look into the possibility of making a low-cost 'people's car'.

The Jha Committee came to the conclusion that it would be possible to make an inexpensive people's car at a target price of around Rs 5,000-6,000 (the equivalent of Rs 1.6 lakhs in 2012), about half the prevailing price of cars then; and that the demand for such a car could well be over 10,000 units per year. Submitted in February 1960, the Jha Committee report recommended the car should be small yet roomy, sturdy, and capable of carrying an average Indian family. The conclusions of the Jha Committee had the Indian government enthused enough for the then minister of industry to announce in Parliament: "We will have the people's car, of the people, manufactured by the people, for the people of this country".

Despite such enthusiasm, in the following year the government decided to create another committee (it was an epoch when one committee begat another), this time under the chairmanship of the retired Chairman of the Railway Board, G Pandey, to look into the feasibility of the 'people's car' project. The Pandey Committee concluded that it was indeed feasible and recommended the car be made in collaboration with French car-maker Renault. The fact that Renault was (French) government-owned may have influenced Pandey's recommendations.

The Renault 4CV — France's people's car from the 1950s.
Renault Corporate Communications/All Rights Reserved

At that point of time, Renault was manufacturing an impressive range of cars, of which the smallest and cheapest was the 4CV, a compact 3.6-metre-long car powered by a small and fuel efficient 747cc four-cylinder engine located at the rear, like the VW Beetle. Unlike the Beetle, however, the 4CV was a more practical four-door car. Going against it was the fact the 4CV had been in production since 1947. As the 4CV was eventually phased out in 1961, this decision must already have been made by the French car-maker in 1960. So what Renault offered instead, was the Dauphine, a small rear-engine four-door saloon that the French car-maker had launched in 1956. Bigger than the 4CV, the Dauphine was markedly more expensive than the 4CV.

The Renault Dauphine, the model offered to the Indian government.
Renault Corporate Communications/All Rights Reserved

That did not seem to have deterred the Government of India, which began discussions with Renault. What did bring an end to the negotiations was the reaction of the then Deputy Chairman of the Planning Commission, V K Krishnamachari, who believed that greater priority needed to be given to bicycles, scooters, buses and trucks. On 9 August 1962, the Minister for Steel and Mines, C Subramaniam, stated in Lok Sabha that, "The small car project cannot be moved up in any list of priorities. The priority in the field of automobiles for some time to come ought to be definitely and overwhelmingly in favour of the manufacturers of commercial vehicles which will provide the base for transport of goods and public transport."

The then chairman of Premier Automobiles, Lalchand Hirachand (step-brother of Walchand Hirachand), had suggested to the Jha Committee and later to the government, that the prices of cars would come down if production volume (which was restricted by the licensing system) of the existing car-makers went up; that it would be unwise to launch a fourth project. This point of view was echoed by the other existing car-makers. True, the numbers for the Indian marketplace weren't all that impressive: though the sales of cars had gone up from a very modest 14,688 in 1950 to 26,800 in 1958, achieving a volume of an additional 10,000 for a fourth car-maker looked a tad ambitious.

With the war with China in 1962 and other matters of greater import coming up, no decision had been taken regarding the 'people's car' project even by 1964. The government, at that point, still was not too clear whether the project would be a public sector venture or a private enterprise. Then came the war with Pakistan in 1965, famine and the foreign exchange crisis of 1966-67. The idea of the 'people's car' project was effectively shelved for the time being.

As the government vacillated whether the project was a priority or not, whether it should be executed by the public sector or private players, and whether it ought to be completely indigenous or foreign, 17 hopefuls had reportedly applied for licences. Among them were the existing car-makers — Hindustan Motors, Premier Automobiles and Standard Motors.

Premier proposed making the Fiat 600D, a 3.29-metre long car launched in 1964, at a price of Rs 9,000, with the promise that the price would come down if volumes went up to 40,000 cars per year. Hindustan, indigenously developed a small two-door car with a rather strange shape and an advanced 'front-wheel-drive' system (it is quite likely that the journalist from the period may have confused 'front-engine' with 'front-wheel-drive').

Standard Motors, the smallest of the three, resorted to a rather ingenious ploy. It took the 6CWT pick-up from Standard Triumph's UK line-up, a back-to-basics vehicle that shared most of its body panels with the Standard

10 that SMPIL was already assembling in India, and developed a shorter, smaller car, with the boot literally chopped short. The one prototype made was dubbed the Stanmobile.

The Stanmobile prototype.

A more serious proposal came from TELCO, which offered to produce a car made by its collaborator from Germany, Daimler-Benz; no, not a Mercedes-Benz model, but a DKW Junior. DKW was a German car and motorcycle manufacturer, founded in 1917 by Jörgen Skafte Rasmussen, a Danish engineer. Originally standing for 'Dampf-Kraft-Wagen', or 'vehicles powered by steam', DKW was soon popularised by the slogan 'Das Kleine Wunder', or 'the little marvel', when it moved to making petrol-engine cars. It was also nicknamed 'Deutscher Kinder Wagen', or 'the German pram'. In 1932, DKW merged with Audi, Horch and Wanderer to form Auto Union. Post-war, as Auto Union struggled to re-establish itself with just the DKW brand, Daimler-Benz took over in 1957.

Launched in 1959, under the Daimler-Benz stewardship, to make inexpensive cars for the mass market, the Junior was a two-door car that, at 3.98 metres long, was actually a little longer than both the Fiat 1100 and the Standard Herald, but markedly lighter than either. Powered by a smaller 796cc two-stroke three-cylinder engine, the DKW would also have been more fuel efficient. Most interesting was the price tag: TELCO said the car would be priced at Rs 6,950 ex-factory, which would have made the DKW significantly cheaper than the three cars then being made in India. TELCO proposed to supplement its truck-making facilities at Jamshedpur with an assembly line for the car, at an expenditure of Rs 14 crore for a capacity of 12,000 units per year.

Of course, TELCO did not get the green signal; neither did most of the other applicants, including the likes of international majors like BMW, British Motor Corporation (BMC), Citroën and Renault from France, Toyo Kogyo and Toyota from Japan, and Germany's Volkswagen. Incidentally, Auto Union had became a part of the Volkswagen Group in 1964, when Daimler-Benz sold it, and beginning 1965, DKWs were badged and sold as Audi, as VW revived the more prestigious brand name. What Volkswagen in turn offered to make in India, is not known but it may well have been the Beetle, which was the hottest selling car across the globe in the 1960s.

The BMW-Isetta bubble car. © Gautam Sen

BMW apparently offered the 600, a slightly bigger, four-seat, in-house derivative of the Isetta bubble car that it had been making under licence. The car was offered to Hindustan Vehicles Limited, a sister concern of Hindustan Motors. In fact, BMW even developed a special version for India, taking into account the peculiarities of Indian weather conditions. Whether the BMW 600, with its odd design, would have been acceptable in the Indian marketplace, is a matter for conjecture.

Another German vehicle manufacturer, Tempo, which had inked a technical collaboration with Bajaj Tempo Ltd (which became Force Motors eventually), in the 1950s, had also developed a micro-car along the lines of the BMW-Isetta, but prettier. It then developed a second prototype that was more car-like, with two doors and the engine located at the rear and the rear wheels further apart, unlike the first proto which had its rear wheels close-set, Isetta-like. This was ready by either late 1956 or early 1957, two years before the Jha Committee and the plans for a 'people's car' took steam, so nothing came of it. Instead, Bajaj Tempo brought out the Tempo Hanseat three-wheeled goods carrier and the Tempo car for India remained a footnote in automotive history, with the sole surviving prototype on display in a museum in Germany.

The 600 – the car BMW offered to Hindustan Vehicles Limited. © Gautam Sen

There may also have been an attempt to make the strange German Fuldamobil micro-car in India, as it was supposed to have been licensed for manufacture in various countries — as the Nobel in Chile and the UK, the Bambi in Argentina, the Bambino in The Netherlands, the Fram King Fulda in Sweden, the Attica and Alta in Greece and the Hans Vahaar in India. There are photos of several Fuldamobils junked somewhere in India — but where exactly, and who could have been the possible licensee, is lost in the mists of time....

The models offered by the other car-makers are not known and we can only speculate on what they may have been. BMC may have offered either the Mini or Austin/Morris 1100 (if the Mini was seen as impractical, given it had just two doors). Price-wise the Mini would have been significantly cheaper than the 1100, probably by some 30 per cent. From Citroën, the guess would be that the 2CV would have been on offer, being its cheapest car, other than the fact that the 2CV epitomised the concept of a 'people's car' more than any other. Or it may have offered the 2CV clone, the Citroën Ami, which was slightly more spacious and modern, even if it was very strange looking (despite which, it was the bestselling car in France then). Into the 1960s, Renault may have offered the 4L, the small and very successful front-wheel-drive model that replaced the 4CV.

The Renault 4L was offered too.
Renault Corporate Communications/All Rights Reserved

Toyo Kogyo, the Japanese car-maker known today as Mazda Motor, launched its first automobile, a three-wheeled goods carrier called the Mazda-Go, in 1931. But the first car from this Japanese company, the Mazda P360 Carol, a compact little four-seater, available in both two and four-door form, was launched as recently as in 1962. A little short of three metres long and powered by a fuel efficient 358cc four-cylinder engine located at the rear, the P360 Carol was Japanese's bestselling small car during the 1960s. It is quite possible that Toyo Kogyo offered the P360 Carol for the Indian small car project, given the price mentioned was a most reasonable Rs 6,700.

Whether Toyota's offering was as appropriate remains debatable as its smallest and cheapest car then was the Publica, a more conventional two-door saloon that was bigger and more expensive than the Mazda, but smaller and more cramped than the Fiat 1100 and Standard Pennant/Herald. Controversial political heavyweight Krishna Menon pushed the case for Toyota. As the incumbent defence minister, Menon suggested the people's car be made at one of the ordnance factories.

Known indigenous efforts were sadly limited to just two: one from Hindustan Aeronautics Limited (HAL), the company created by Walchand Hirachand but then taken over by the government; the other a creative effort by a certain Shankarrao Kulkarni from the small town of Ichalkaranji, near Kolhapur, in Maharashtra.

The HAL initiative was the work of P M Reddy, a general manager at HAL, who developed a car in-house in 1964. Unfortunately, information remains sketchy and details of the specifications, size and type of car have been lost in the mists of time. But a photograph of the car would seem to indicate the design was rather derivative — bits and pieces seem to have come from other cars and it is quite possible the HAL car was developed from an existing one, but well disguised.

That may very well have been the case with the other indigenous effort too, the Meera. Shankarrao Kulkarni of Ichalkaranji developed a tiny car sometime in the 1960s, though in some reports it has been said that the first prototype was made in 1949. But this is doubtful as the design is more typical of the models from the late 1950s or early 1960s. Karl Bhote, an automotive enthusiast and expert on Indian automotive history, specialising specifically in period registration numbers, believes the number plate on the prototype Meera indicate the car could have been from any time after 1962 and before 1972; this has been confirmed by Cedric Sabine, who looks after the Indian section of the number plate histories club, Europlate.

The Meera was significantly smaller than the cars made by India's Small Three, with the engine located at the rear. According to a leaflet, the car was developed by Shankarrao Kulkarni, 'very famous in engine designing'. Apparently, Kulkarni undertook 'an industrial tour in European countries, in Germany, and visited many motor industries as well as exhibitions'. Incidentally, Shankarrao Kulkarni worked for industrialist Kirloskar and had as many as 10-15 people (one of them being a certain Vasudev Deshpande), working on different parts of the car. Components came from parts manufacturers such as Ceat Tyres, who supplied the tyres, Lucas provided electrical parts and the glass bits came from Ogale Glassworks. Kulkarni's grandson, Hemant Kulkarni, has been reported to have said that his grandfather studied till just standard seven. But in automotive history there have been several cars designed by people who were not engineers.

Kulkarni made six prototypes (or six evolutions of the prototype?), the last in 1970. By then the styling had evolved to a rather distinctive, sharp-edged, boxy look. Interestingly, these (presumably very tight) four-seat cars featured three doors: two at the front, like all two-door cars, plus an additional one on the left, to give access to the rear bench seat — an idea Kulkarni may have got from the BMW 600, which, for rear seat access had a single door on the right (as one drives on the right in Germany, so the kerb is on the right, unlike in India, where we are supposed to drive on the left of the road).

With its overall length at a little under three metres (2.97 metres exactly), and both width and height at about 1.3 metres, tiny 10-inch wheels, along with motive power from a two-cylinder engine developing a claimed 14bhp, the Meera was (in dimensions and specifications), surprisingly similar to several German bubble-cars of the 1950s, most specifically the Goggomobil. Though the design of the body does not match the Goggomobil's, in overall profile the Meera could very well have been inspired by it. It is quite possible that during Kulkarni's extensive travel through Europe, especially Germany, he had seen the Goggomobil, one of Germany's bestselling micro-cars during the 1950s and 1960s. It is also possible Kulkarni may have come across a Goggomobil or two in India.

The Goggomobil was made by Hans Glas GmbH. Traditionally a manufacturer of sowing and farming machinery, Hans Glas from Dingolfing, in Germany, had turned to car and scooter manufacturing after WWII. Under the stewardship of Andreas Glas, in addition to a Vespa-inspired scooter named Glas Goggo, a range of cars was developed; the tiny bubble-cars that Hans Glas GmbH launched carried the Goggomobil branding. The Glas branding was also used for the bigger machines.

Made from 1955 to 1969, the 2.9-metre long Goggomobil featured a 14hp 293cc two-cylinder engine located at the rear. The two-door four-seat car saw success not just in Germany and several other European countries, but was also exported to the US (with a bigger and more powerful 392cc engine developing 20bhp and named the T400). In Australia, the Goggomobil was produced under licence and a pretty little sports two-seater was made in Sydney by an Australian company called Buckle Motors.

Harit Trivedi, a collector of cars and automobilia, more specifically vintage literature and photographs of cars and vehicles from India, speculates that if the Goggomobil had seen success in Australia it would have been natural selection for India. In December 2012, Harit came across conclusive evidence that the Goggomobil was one of the cars that was under consideration for India's 'people's car'.

CHASSIS & CHASSIS DIMENSIONS.

Wheels	..	Steel disc
Number of wheels	..	4 and 1 spare
Tyres	..	4.40'-10"
Inflation pressure front	..	16.2 lb.
Inflation pressure rear	..	23.5 lb.
Front axle	..	Swing axle, with helical springs
Rear axle	..	Swing axle, with helical springs
Shock absorbers	..	Double-acting, hydraulic

Dimensions :

Wheel base	..	70.8 in.
Track	..	Front: 3' 4 1/8", rear: 3'-3¾"
Ground clearance	..	8"
Type of design	..	Chassis-cum-floor of body of sheet steel construction.
Admissible total laden weight.		14.3 cwt.
Empty weight	..	8.2 cwt.

The world's only Small Car with the big Car Suspension

PERFORMANCE :

Top Speed	..	62 m.p.h.
Max. gradient	..	36%
Fuel Consumption	..	67.2 Miles per (Br gallon) car
	..	58.8 Miles per (Br gallon) van

For further particulars please write to :

ACI

Head Office,
Commerce House, Currimbhoy Road, Ballard Estate, Bombay 1.
Telephone No. 263661-2-3.

OR

Branches at

73-A, Ganesh Chandra Avenue, Calcutta 13. Telephone No. 242058

17-18, Indira Palace, 2nd floor, Connaught Place, New Delhi 1. Telephone No. 43088

Agurchand Mansions, 1st floor, 35, Mount Road, Madras 2. Telephone No. 84278

National Blocks Bombay

GOGO

LITTLE CAR FOR THE LARGE FAMILY

A. C. I. Introduces the ideal small car which can be purchased by a man of average means and therefore best suited for middle class people in India.

The Government of India had appointed a special Committee under the Chairmanship of Shri L. K. Jha to select the most suited Car in this category from the several offers submitted to the Government.

Messrs. Associated Corp. of Industries (India) Private Ltd; Bombay. had far back as early july 1959 submitted to the Government of India, a manufacturing programme for the GOGGOMOBIL, one of the most remarkable Small Cars. She is the very essence of beauty wedded to common-sense. One look and youwill like her enormously — What else could make sense ,than to power this design with the sturdiest, most dependable GLAS engine. One drive in her and you will go for the GOGO.

Nowhere else in the world of small automobiles can you find such a happy combination. The Goggomobil Saloon has also a sister now, a Pick-up Van, besides other models within the price range.

Now step back and gaze upon the latest 1960 model - A four-seater with a new 400 CC engine - excellent independent suspension-rapid acceleration - superb roadholding - low maintenance costs - robust chassis - easy manoeuvrability-with a ground clearance of 8".

The GOGGOMOBIL is a small car which has been designed and built by HANS GLAS, West Germany in just recognition of the need and demand of such a car at this time, when we ask for low running costs and great fuel economy and yet are not willing to compromise excellent driving qualities.

The Goggomobil sales leaflet. Courtesy Harit Trivedi

A sales leaflet that Trivedi acquired in 2012 explains that a certain Bombay-based company, Associated Corporation of Industries (India) Private Limited (ACI) had as "far back as July 1959 submitted to the Government of India a manufacturing programme" for the Goggomobil. Featuring the US-spec engine (which gave the car a top speed of 100km/h), the Goggomobil was going to be priced at "about Rs 5,000, ex-factory, when progressively manufactured in India, excluding taxes" claimed the sales leaflet. Along with the India-specific leaflet, Harit also picked up some two dozen sales brochures of the Goggomobil range — comprising the 682cc Goggomobil Royal (also sold as the Glas Isar T700), the convertible and coupé versions of the Goggomobil and the van.

Whether these were also planned for manufacture is not known, but the Goggomobil T300 or T400 saloon surely was, as Harit also came across three photographs of a Goggomobil with a temporary Delhi registration plate: one shot is of the car on the street, another with Morarji Desai (India's home minister during the early 1960s), and yet another — and the most telling — one with India's then Prime Minister, Jawaharlal Nehru.

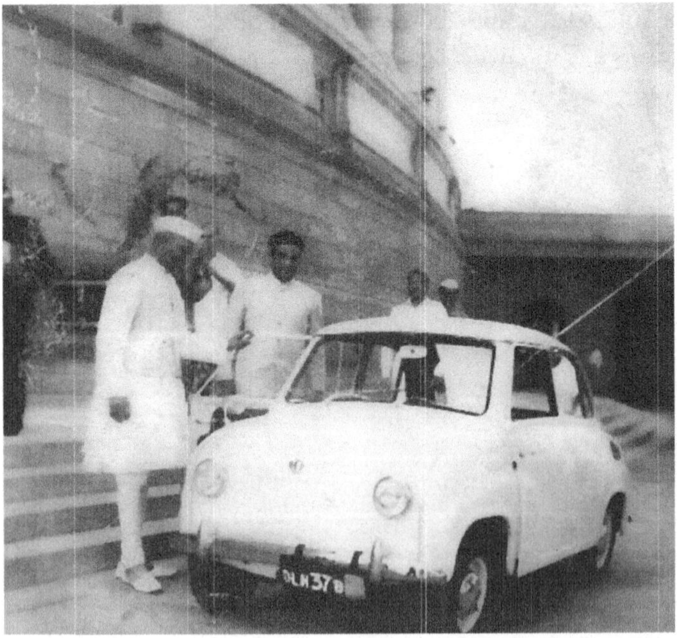

Pandit Nehru and the Goggomobil. Courtesy Harit Trivedi

Clearly, the Goggomobil was shown to the most important people in the Indian government, ACI appointed dealers in Delhi, Calcutta and Madras, and set up headquarters and operations in Bombay. We can safely speculate that a great deal of lobbying must have been resorted to. But obviously nothing came of it. Whether the one right-hand-drive Goggomobil that came into India went back to Germany, or just disappeared, or morphed into some 'indigenous' proposal, is not known. All we do know is that Hans Glas GmbH became part of BMW in 1966, when the latter acquired the former as it needed to expand production capacities. ACI did not get the licence to make the people's car, nor did Shankarrao Kulkarni, nor TELCO, nor HAL, nor for the matter, any of the international majors.

That privilege was accorded to a certain Madan Mohan Rao and a certain Sanjay Gandhi.

On 13 November 1968, the Minister of State for Industrial Development, Raghunath Reddi, announced in the Lok Sabha that his ministry had received a proposal to make an inexpensive people's car from Sanjay Gandhi, son of the then Prime Minister, Indira Gandhi. The proposal from 21-year-old Sanjay Gandhi was for a small car with a top speed of 85km/h, run more than 18km per litre, yet cost just Rs 6,000. Reddi also mentioned that the government had received proposals from 13 other car-makers, entrepreneurs and engineers.

Till November 1970, there was talk that the people's car would be made by the public sector. The Union Cabinet held discussions with several international majors, of which the most noteworthy was Renault, as the state-owned French major offered to cover the cost of foreign exchange. Yet the government decided the investment involved was considerable and the best interests of the nation would be served by having the people's car made by the private sector. On 31 November 1970, the Minister of Industries, Dinesh Singh, handed over a letter of intent to Sanjay Gandhi, to make the people's car. Apparently the decision had been taken at a Cabinet meeting presided over by the Prime Minister herself.

Sure enough, controversy flared. The first point to come up was Sanjay Gandhi's qualifications: what special abilities did he have to be awarded such an exclusive licence, other than being the Prime Minister's son?

ENTER SANJAY GANDHI

Sanjay Gandhi was a diehard car fan. Prominent journalist and writer Vinod Mehta, who wrote a book on Sanjay Gandhi in 1978, believes that it was his father Feroze Gandhi's obsession about tending his immaculately maintained Morris that profoundly influenced the impressionable young boy, giving rise to his passion for automobiles. Whenever Sanjay spent time with his father — who was estranged from his mother when Sanjay was still very young — he observed him polishing, cleaning and repairing his car constantly. At a young age Sanjay expressed his desire to make cars the vocation of his life — an ambition that even his illustrious grandfather believed his grandson had immense talent for.

Sanjay was never academically strong at school and was reluctant to pursue further studies. His mother, given her connections, managed to finagle a prestigious three-year apprenticeship for the young boy at no less a car-making venture than Rolls-Royce Motors, starting September 1964. The apprenticeship was designed to train technicians for semi-skilled and skilled jobs at Rolls-Royce's demanding assembly line; not the equivalent of an engineering degree. Yet talented apprentices could find themselves in the prestigious design and engineering department of Rolls-Royce once they completed their tenure.

Sanjay Gandhi did not manage to stick out the three years of apprenticeship, leaving at the end of two years as he believed he had learnt all that he needed and there was nothing more to be gained by staying on at Rolls-Royce. Whilst in the UK, Sanjay also made newspaper headlines for the wrong reasons. In December 1966, he was arrested for driving without a valid licence. He also managed to have a couple of smash-ups with his Jaguar XK120 (showing good taste but poor driving ability). Not particularly popular with his fellow apprentices, Sanjay was not missed when he left Rolls-Royce to head back to India.

Back home, Sanjay got down to his automotive obsession — to design and develop a small car for the masses and provide the cheapest set of wheels to the nation. Sometime in 1967, Sanjay rented a shed in Motia Khan, a semi-industrial complex in the heart of Old Delhi, adjacent to one of the biggest wholesale markets of Asia — the Sadar Bazar. Here he set up a small workshop where he employed some craftsmen, including denters (metal-body workers), car mechanics, lathe machinists, welders, painters, electricians and a hands-on man who would oversee the entire operation. As he added to his team, he moved to larger premises at Gulabi Bagh.

Work began on the development of 'Sanjay's car'. The usual process of designing and developing a car in those pre-computer days started with sketches and renderings, which were then converted into detailed scale drawings, from which 1:4 or 1:5 scale models would be developed. Once the shape and style had been finalised, full-size scale drawings would follow, along with packaging studies to ascertain the habitability of the car. Then wooden forming bucks would be made, from which the metal panels for the body of the prototype would be beaten out. The chassis would be fabricated using jigs and fixtures. Whether Sanjay Gandhi followed such a system is doubtful, given the garage-like set-up the young man had.

When, in November 1968, Sanjay Gandhi submitted his proposal to the Indian government, he did not have any car to show. Whether extensive paper drawings and presentations were made to make his case is not known but seems to have been most unlikely. The project report Sanjay Gandhi submitted to the Government for scrutiny envisaged a two-door car powered by a two-stroke, two-cylinder, rear-mounted engine developing 14bhp, with 'tyres similar in size to scooter rickshaw tyres', which sounds suspiciously similar in specs to the Goggomobil. The project report also claimed that no foreign collaboration or import of machinery, raw material or components was envisaged.

Two years later, when Sanjay Gandhi was granted his letter of intent, his car was still not ready. It is worth noting though that letters of intent were granted to ten others too, one of them being the aforementioned Madan Mohan Rao. At the same time, the government put up stringent requirements: foreign collaborations, consultancies and import of capital goods and components were not allowed. And before the letter of intent was converted into a licence, prototypes had to be developed and would need to be tested and approved. Given the kind of hold Sanjay was reputed to have over his mother and Indira's tendency to take unilateral decisions, it is possible these conditions were imposed on the insistence of Sanjay, with the sole purpose of eliminating competition. This, as we will see later, worked as a double-edged sword.

Madan Mohan Rao's letter of intent, along with those of the other contenders, lapsed because they could not fulfill the conditions imposed by the Government. Sanjay Gandhi presented his prototype small car in November 1972. The car was named Maruti, one of many names of the Hindu monkey god, Hanuman. A little over a year earlier, on 4 June 1971, Maruti Motors Limited had been incorporated. The unveiling of Maruti was quite an event. The people's car, of the people, manufactured by the people, for the people of this country, had finally arrived. Or had it? The price had jumped to Rs 11,300, two-thirds more than the original proposed price of Rs 6,693. The increase in raw material cost was touted as the reason.

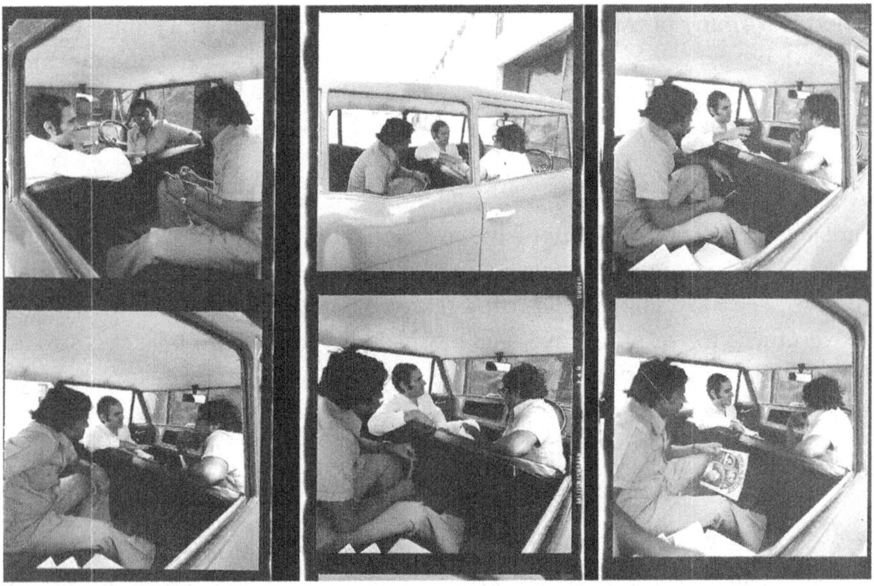

Sanjay Gandhi being interviewed in his car. From the author's archives

India's first go at an indigenous people's car was not unattractive. The shape, to say the least, was rather rudimentary and boxy, with hardly any finesse in the design; yet it was not exactly ugly. Adopting a two-box look, the car was rounded from the front, borrowing quite a bit from the 1959 Mini, but with a shorter bonnet. It ended with an abrupt, almost vertical rear. The absolutely flat front windscreen had minimal rake, and the design was hardly aerodynamic in its form. The beltline kicked up into an interesting hump at shoulder point.

Top view of Sanjay Gandhi's Maruti. From the author's archives.

Sanjay's Maruti from the front, which appears Mini-like. From the author's archives.

Sanjay's Maruti was flat and featureless from the rear. From the author's archives.

The car used a whole host of proprietary parts from other popular cars of the period, specifically the older generation Fiats and Ambassador. Door handles came from the Fiat 1100 of the 1950s, door locks, quarter glass frames, dashboard metres, headlamp assemblies, all came from the Ambassador. Delhi-based restorer Tutu Dhawan, who was familiar with the car as he had one till the 1990s in his garage, believes the floorpan of the car was derived from the Ambassador and modified to suit the shorter wheelbase. The front suspension system, featuring wishbones with torsion bar, was also strikingly similar to the Ambassador's. The car did not have a front grille as the air-cooled engine was at the rear. The sloping, rounded front featured headlamps that came from the Ambassador, and the chromium-plated steel rings were from the Enfield Bullet.

The car was a two-door, four-seater. The front seats featured folding backrests, facilitating ingress to the rear bench seat. But getting in and out for the elderly was not comfortable. Yet, passenger volume, given the car was significantly smaller than what was selling in India then, was impressive. Small 10-inch wheels saved space inside.

Thin wraparound front and rear chrome-plated bumpers made from mild steel provided visual relief to a form that was essentially featureless. The body sides were almost flat, with slight contours defined by parallel swage lines, to give some strength. The two doors had flat, winding glasses, but the windows at the rear were fixed, though they featured folding quarter glasses aft. Seats had well-sprung cushions with rexine doing the job for the headliner. Ordinary rubber mats covered the thinly insulated floor. Sound deadening or insulating material was not used at all, essentially to keep costs down. The promise was to sell the car much cheaper than the Ambassador, Fiat 1100 (Premier President) and Standard Herald.

On paper, the concept of the vehicle was rather original. The shape and packaging took a lot from the Mini school of thought in featuring a 'two-box' look that was still new in the 1960s. Yet the styling was already dated for its time as automotive design trends by the end of the 1960s, was towards the sharper-edged origami look. In terms of mechanical configuration too, Sanjay Gandhi chose something that was getting rapidly outmoded. Most car-makers, including Fiat, Renault and Volkswagen, who had been the greatest exponents of the rear-engine cause, were moving towards front-engine front-wheel-drive layouts. Yet Sanjay chose to go the old-fashioned way.

Sanjay's Maruti under 'construction'. From the author's archives.

True, in some ways, the rear-engine layout was easier and cheaper to make and that thinking may have influenced him. Or it just may be possible that he had access to such a mechanical configuration when he was developing his car, as Dhawan suspects, conjecturing that the entire rear mechanical elements came from an existing automobile. Harit Trivedi too, is convinced the Maruti was derived from the Goggomobil. Indeed, from the front, the Maruti does have more than a striking resemblance to the Goggomobil.

But the engine of the Maruti was a horizontally-opposed twin, similar to that of a BMW motorcycle, and in that sense, different from the in-line two-cylinder of the Goggomobil. Dhawan also points out that all the stampings and markings on the engine had been rubbed out.

The location of the engine, gearbox and differential were all along the rear axle of the Maruti, whereas most rear-engine cars, including the Goggomobil, had the powerpack 'overhung' at the rear, aft of the rear axle. Also, Sanjay's Maruti was impressively wide. Dhawan estimates it at significantly more than 1.4 metres, whereas the Goggomobil was fairly narrow at under 1.3 metres.

Either way, it seems that Sanjay Gandhi and his merry bunch of mechanics designed the car without any external help or advice. As the Opposition questioned the Government's decision and accused the Prime Minister of nepotism, the Government decided that Sanjay Gandhi's Maruti Motors would get its licence only after the Maruti prototype passed certain types of tests. A committee consisting of the representatives of the DGTD (Director General of Technical Development), the VRDE (Vehicle Research and Development Establishment) and the Ministry of Heavy Industries and Public Enterprises (with representatives from Maruti Motors present), prescribed that the car should run for 30,000kms, and after a reliability test of 10,000kms, a view should be taken regarding the licence.

Apparently the prototype did complete 10,000kms and the test report showed 'satisfactory' performance on important counts. The Letter of Intent was converted into an industrial licence by the Ministry of Heavy Industries after the VRDE had been informed. On 30 March 1974, VRDE issued Maruti a clearance certificate. Though the prototype did cover 10,000kms, it did not cover the recommended 30,000kms. In fact, testing and development of prototypes is perhaps the single most important aspect of designing and developing an all-new car for series production. Major car-makers test and run scores of cars over millions of kilometres, always learning, modifying and reworking the smallest details before signing off the definitive design. Maruti Motors did not do so with the original Maruti.

In all fairness, Maruti Motors had very limited resources. It began with an authorised capital of Rs 2.5 crore, which went up to Rs 10 crore by June 1972. But the total paid-up capital by September 1974, was Rs 1.8 crore, whereas the value of the net assets of the company as on 31 March 1974, was Rs 4.5 crore. In 1972, Maruti Motors began appointing dealers, from whom deposits were taken. Rs 2.18 crore was collected from 73 dealers.

The basic Maruti factory with virtually no equipment. From the author's archives.

Yet Maruti Motors' investments were not all that substantial. The 296.7 acres of land was controversially acquired for just Rs 36 lakh, with farmers being unceremoniously evicted, thanks to the efforts of the Haryana Government, led by Chief Minister Bansi Lal, who was bending over backwards to please the Gandhi family. Looking at photographs of the factory from that period it is obvious that machinery and equipment were minimal. There was not even a conveyor belt for the assembly line; the cars being pushed manually from one work station to another on trolleys.

Few Marutis were produced at this time. From the author's archives.

Vinod Mehta, in his book on Sanjay Gandhi, quotes from the *Sunday Times* (London) report on the Maruti factory from April 1977: 'The body shop where the workers apparently hand-stitched the little Maruti bodies is a joke. So also is the engine shop, the pattern shop and the foundry, and the whole place with three ground level fireplaces looks like a dirty indoor barbecue.' Mehta estimates that total investment could not have topped Rs 6 crore — for a project that needed at least ten times as much. When the Government had discussed a public sector 'people's car' project with Renault and other international majors, the estimated cost was Rs 57 crore.

But underfunding was not the main problem. The bigger problem was Sanjay Gandhi's attitude. With an incomplete Rolls-Royce apprenticeship as his only qualification and work experience, Sanjay believed he knew all he needed to know about car-making. In 1971, he went to Europe and toured several car manufacturing ventures and came away convinced there was nothing he did not already know about the auto business.

When the first Maruti prototype was unveiled in November 1972, Sanjay promised that the first commercial production cars would roll out just five months later, by April 1973. In 1974, after Maruti received its VRDE certification, the car was claimed to be at 'the threshold of production'.

In the 25 May 1975 issue of *The Illustrated Weekly of India*, veteran editor Khushwant Singh, wrote about his visit to the Maruti factory, where he went for a short spin in one of the pre-production cars: 'I take the wheel. Sanjay sits beside me. The gear is on the floor. As I release the clutch, it lurches forward. Although it has only two cylinders and 8.3 horse-power, it responds to the pressure of the accelerator like a more powerful car. Its steering is also more sensitive to the touch and makes the wheels turn with greater alacrity than the steering-wheel of any car I've driven. Despite its minuscule size, it holds the road very well and attains its maximum speed of 110 km per hour without overstretching itself. I take the 60-degree bends at 60 kilometres per hour with the greatest of ease. My fear that it has too low a clearance for the rough Indian roads is soon dispelled. Sanjay takes over himself and drives it at breakneck speed over bush, brier and mound.'

A great fan and defender of Sanjay Gandhi, Khushwant Singh nevertheless wondered: 'How the Maruti will last remains to be seen'. He also succinctly pointed out that the car was no longer a people's car 'because it will cost Rs 25,000 (ex-factory Rs 16,500; in Haryana Rs 21,000). Nevertheless, it will be Rs 5-10,000 less than any of the three Indian cars in the market.' Overall,

Khushwant Singh was positive: 'Soon little Marutis should be seen on the roads of Haryana and Delhi; and a month or two later they will be running between Kalimpong and Kanyakumari.'

But that did not happen — not with Sanjay Gandhi's Maruti.

Emergency was imposed in June 1975. Sanjay Gandhi had other matters to attend to. In the meantime, to generate some business, another entity, Maruti Heavy Vehicles, was constituted to make bus bodies and road rollers. During the Emergency, business was good for Maruti Heavy Vehicles, but with the Emergency coming to an end, business collapsed. With the Janata Party coalition being voted to power, a series of enquiries against the Gandhi family and Maruti ensued. When, on 6 March 1978, the High Court of Punjab and Haryana ordered the winding up of Maruti Motors, not more than 40 Marutis, at most, had been made.

In January 1980, the Congress party came back to power and Indira Gandhi was back in the driving seat. Apparently, Sanjay was excited about the possibility of reviving Maruti. But fate had other plans. On 23 June 1980, Sanjay Gandhi crashed whilst flying a Pitts aircraft, and was killed instantly.

India's first small car project had a lot of promise, and given the kind of proximity it had to the powers that be, the venture should never have failed. But making a car is not easy. Making a body and giving it wheels was the easier part of the job, as most of the bits and pieces came from other automobiles being made in India or elsewhere. But putting that into series production was another matter.

And then there was this little matter of the power pack. For the pre-production prototypes Sanjay resorted to using imported motorcycle engines that were matched to a four-speed gearbox, which must also have been sourced from somewhere. All very well for pre-production cars, but what about production cars? That is when the project struck a roadblock, specifically the power pack. The understanding was that the Indian 'people's car' was to be built indigenously with all major and minor components of the vehicle being produced in the country. It was on this premise that Sanjay Gandhi was the only one to receive a new licence to manufacture cars. But Sanjay and his inept team did not have the ability or know-how to duplicate the engine for series production. Sanjay did get a German consultant who reportedly brought in a couple of engines, but the consultant soon realised that neither Sanjay nor his team had the ability to copy and bring the power pack into production.

Several senior people in the Indian automobile industry opined that the conditions imposed by the Letter of Intent about indigenous design and locally-sourced equipment, made it well-nigh impossible for Maruti to succeed. Though Maruti was not allowed to import any machinery, it could have used the sophisticated machine tools HMT was manufacturing then. Maruti could have also got machinery from East European countries such as Czechoslovakia (which had a fairly progressive car industry then, manufacturing the Skoda and Tatra), against rupee payment under the Stock & Trade arrangement that India had with Eastern Bloc countries.

In a rather defensive article on Sanjay Gandhi and the Maruti mess, in the 11 March 1979 issue of *The Illustrated Weekly of India*, T A Pai, the then Minister of Heavy Industries, points out: 'The disbelief that we could do it in India on our own is the usual belief prevailing in the country and in the Government and the financial institutions which has always resulted in many of our people not being able to utilise their ideas, their ingenuity and their resourcefulness for development in the country.'

Without doubt, Sanjay Gandhi, or for the matter anyone with automotive knowledge, had every right to experiment and try to develop a car for the people. Given the right opportunity and appropriate resources, this writer believes that several brilliant Indians could have developed an indigenous people's car that would have worked. True, by the 1960s, the automobile industry was fairly advanced; yet there were several budding, young car-makers from many countries across the globe, including Japan, who were just beginning to cut their teeth in the car-making business, developing brand new cars in-house.

The problem with Maruti was Sanjay Gandhi. A two-year apprenticeship was insufficient; a simple fact that Sanjay Gandhi, his mother and all her cohorts, refused to acknowledge. Sanjay's competence was no greater than that of a decent mechanic at best. Designing and developing a car calls for excellent engineering ability, great design skills, very thorough testing and developmental integrity, plus ample resources and very deep pockets. Sanjay's arrogance, overconfidence and hubris did not help matters either. T A Pai also believes that Sanjay was eventually not serious about the project: 'Once Emergency was declared, the project did not make any progress as he shifted his activities to other fields.'

An all-powerful Sanjay Gandhi during the Emergency years could have used his position to have his car readied for series production. Instead, he was more bothered about dealing with his detractors. Pai, in his *Weekly* article, points out that those who were opposed to the Maruti project got into serious difficulties when Emergency was imposed: 'If Sanjay could have pursued the car successfully, Emergency would not have possibly come.' Failure may have driven Sanjay Gandhi to press his mother into taking the extreme steps she did in fact take, with disastrous consequences to the country.

Yet, India recovered and got back her democracy. Indira recovered and got back her Prime Ministership. And Maruti recovered, or more appropriately, was re-incarnated, and India got back its 'people's car'.

Chapter IV
The Reincarnation of Maruti

With the death of Sanjay Gandhi, Indira was devastated. Though she had lost her favourite son, it was still possible to keep alive his dream. So, soon after Sanjay's death, Indira Gandhi called Arun Nehru, the son of a cousin, and a professional manager heading Calcutta-based paint company Jenson & Nicholson, and asked his opinion on how the 'people's car' idea could be revived.

Arun Nehru went through all the papers and documents and came to the conclusion that, yes, the idea of a 'people's car' was still viable and that it would be possible to revive Maruti. But the only way the project would work would be if the car was to be developed or made in collaboration with an international major, using foreign technology. And to be viable, production capacity needed to be at the least 100,000 cars per year.

Fairly rapidly, the assets of Maruti Motors Limited were acquired by the Government and a 100 per cent Government-owned Maruti Udyog Limited was incorporated on 24 February 1981. It is quite possible that '..if Maruti hadn't been nationalised, the liquidation proceedings would have brought many skeletons tumbling out of the cupboard, and might have even resulted in the prosecution of the guilty individuals', says a report on Maruti published in the 12-25 April 1982 issue of *Business India* magazine. Writers Dilip Cherian and Mohan Ram added: 'Moreover, Indira Gandhi's sentiments in completing her son's dreams cannot be forgotten. The government's decision appears to have had more to do with maternal psychology than with hard headed economics.'

However, the *raison d'être* the Government gave was: 'Looking at the present state of the car industry in the country, Government felt that it would be desirable to set up a public sector undertaking under the Central Government for the manufacture of passenger cars and commercial vehicles. The purpose of nationalisation was modernising the automobile industry, effecting more economical use of scarce fuel and ensuring a higher production of vehicles.'

So, within a little over a decade, from the buzzwords of *swadeshi*, self-sufficiency, indigenous technology and the saving of precious foreign exchange, the Government made a 180-degree U-turn. Foreign technology became a must, along with the need for a joint-venture partner; the issue of foreign exchange was no longer important (except for a rider that the joint venture partner needed to promise a buy-back of vehicles). So the 'people's car', to be manufactured by the people, for the people of this country, was going to come from some other people in some distant land.

The objectives had changed. The means to get to those objectives had also changed, but not the name. The new company, very importantly, still used the Maruti moniker, a tribute to the memory of Sanjay Gandhi. Even if Sanjay's Maruti had not succeeded, a Maruti would be made one day.

Importing a product and factory to establish a vast national car project was not new. Countries like Spain, Yugoslavia and Poland had based their national car industries on imported Fiat technology. The USSR too, followed suit after an attempt at indigenously developing an automobile industry with homegrown products such as Moskvitch and Volga did not quite succeed. The mighty Russians went to little Italy's Fiat for technology and expertise, transplanting the Fiat 124 and its manufacturing facilities to the city of Togliattigrad in south-western Russia. Between 1969, when production commenced, to 1979, over five million Lada (as the car was branded), had already been sold, before the car became an even bigger sales success internationally as one of the cheapest automobiles on sale in the developed markets of Europe, Canada and Australia, earning the Soviet Union much needed foreign exchange.

Though countries like Japan and South Korea had followed (or were still following, in the case of the latter), an altogether different route for their nascent automobile industries, for Indira Gandhi and the mandarins in Delhi, Arun Nehru's suggestion of a people's car 'acquired' from an existing car-maker seemed to be the safest route. To take that route, the first essential was to identify the product and a foreign car-maker who would willingly part with technology and participate in the making of India's 'people's car'. But even before that, a core team to head the project needed to be established.

The choice of heading the project fell on V Krishnamurthy, who had earlier turned around the unwieldy public sector company Bharat Heavy Electricals Limited (BHEL). Krishnamurthy was appointed Vice-Chairman cum Managing Director. For the Chairman's post, Indira Gandhi zeroed in on Sumant Moolgaokar, the legendary chairman of TELCO. Getting the head of a very successful private truck-making venture to be non-executive chairman to India's people's car project made very good sense and gave the project much-needed credibility. Krishnamurthy in turn handpicked D S Gupta, an Indian Institute of Management, Ahmedabad alumnus, who had worked under him at BHEL. Soon thereafter, R C Bhargava, who had been the director for commercial affairs at BHEL, on deputation from the Indian Administrative Services, joined the team. With the core triumvirate at Maruti Udyog in place, work began in earnest to identify companies for partnership and products that would be appropriate.

Though two of the triumvirate had excellent experience managing giant enterprises, and more importantly, dealing with Government bureaucracy, none of them had a clue about the world of cars, which in the early 1980s, was an intricate politico-industrial complex. The oil crisis of 1973 and 1979, along with the Iran crisis, had changed the automotive world order drastically. In 1980, the US was overthrown as the automotive world's top dog (for the first time since 1904, when the US had wrested the crown from France). Japan became the leading car-making nation in 1980, producing over seven million of the approximately 30 million cars produced globally; the US in second place made 6.38 million, while the UK, which used to be the world's second largest car-maker till 1960, dropped to seventh place, manufacturing less than a million cars. Continental Europe produced more than 11 million cars, with Germany and France producing a little over 3.5 million and just under three million respectively. Fiat made most of the 1.45 million cars made in Italy while Spain produced over a million cars.

For the sake of perspective, it is worth noting that in 1980, India and South Korea made 30,538 and 55,000 cars respectively; China's auto manufacture was negligible.

THE SEARCH FOR A PARTNER

With Europe looking like the automotive centre of the world, the Maruti triumvirate decided that it made better sense to look at collaborating with a European car-maker than an American one, given that two of the Big Three were bleeding badly and one needing a Government-backed bailout. Also, the US car-makers made mostly big gas guzzlers, cars that were just the opposite of a frugal 'people's car'.

Before Krishnamurthy, Bhargava and Gupta could decide which and what, they faced some not-so-subtle pressure from several quarters pushing them in the direction of Renault and the Renault 18 model. The government had imposed a rider that the joint-venture partner, other than taking a 40 per cent stake, had to buy back 50 per cent of production. Unlike most others, Renault was fine with the arrangement but insisted it would buy back (the model being the pick-up version of the Renault 18), to sell in the Asian market, at a price below the FOB price of the kits coming into India.

The Renault 18, the car that Renault proposed to the government.

The three met with Renault and explained that the buy-back economics just did not make sense and that the Renault 18, an Ambassador-sized car, was just too big and expensive and would not fit the idea of a reasonably priced 'people's car'. In all fairness, the Government, at the early stages of the project, had recommended that India's people's car should be a six-seater. Though the Renault 18 was a five-seater, it could be configured to take three in the front row, with modified seats. A market survey conducted by the Indian Market Research Bureau (IMRB) on Bhargava's suggestion, showed there was significant potential in India for a small, reliable, low priced car. "Over 90 per cent of car usage in India was with less than four passengers; over 90 per cent of the usage was within the city and not for outstation trips," wrote Bhargava in his excellent book, *The Maruti Story*.

So Renault offered the smaller Renault 5, a very successful small car that had sold in the millions in France and across Europe, and even in the US, a car that perhaps epitomised the idea of the 'people's car' for Europe in the 1970s, better than any other. But as the model dated from 1972, the three decided against it. Renault then offered the 11, a four-door (five-door in car lingo, with the hatchback tailgate as the fifth) launched in 1981, smaller than the 18 and priced halfway between the 5 and the 18. But that did not make it significantly cheaper than the cars selling in India at that point of time.

The hugely successful Renault 5 (in five-door hatchback form) was offered to Maruti Udyog. This is the 3-door version. Renault Corporate Communications/All Rights Reserved

The Renault 11 which was offered as an alternative to the older 5.
Renault Corporate Communications/All Rights Reserved

The Renault Super 5 was a very modern car for the 1980s. This is the three-door model; there was a five-door too. Renault Corporate Communications/All Rights Reserved

But the replacement of the 5, the Super 5, that Renault was developing for launch in 1985, was not disclosed to the Maruti representatives. Rival French giant Peugeot was more forthcoming. Peugeot offered the new small car it was then developing. The project was named internally as the M-24, and later launched in 1983 as the Peugeot 205. Acknowledged by many as one of the best small cars in modern times, and named the Car of the Decade by respected British magazine CAR, the Peugeot 205 would have made excellent sense as India's 'people's car'.

But there was a problem. The Peugeot 205 was to be launched in France in 1983; the earliest it could have been assembled in India would have been late 1984. But Indira Gandhi had given the Maruti core team a target date of rolling out the first Marutis by end-1983.

The Coupé Peugeot 205, the excellent car Maruti missed out on. © Gautam Sen

Bhargava mentions in his book that Peugeot, 'till that time was only producing large cars'. The Maruti team, sadly, did not seem to be aware that Peugeot were also making a small car — the very successful Peugeot 104. But the

104, which had been around since 1972, could have been seen as too dated. The Maruti management also did not seem to be aware that Peugeot owned Citroën at that point of time — a company it had taken over in 1974, when the latter was in serious financial difficulties.

Citroën had in its range the Visa, a spacious five-door five-seater, launched in 1978, that was not only cheaper than the Peugeot 205, but also featured amazing ride quality akin to a limousine, with soft, long-travel, fully independent suspension. The rear seats were split in the centre, folding individually, and could be removed altogether, providing cavernous load-carrying capability. A single wiper was sufficient to sweep the deep and curved windscreen and the design was impressively aerodynamic for its time, making the base 652cc two cylinder-engine version of the Visa a very fuel efficient car. A sort of Citroën 2CV for the 1970s and 1980s, the Visa, a true grandchild of the legendary 2CV, sadly got missed out in the search for India's people's car.

So did the Fiat Uno, another great small car launched in 1983. The Maruti core team had approached Fiat, which already had a reputation for making excellent 'people's cars' with its Topolino and Fiat 500. But at the time Maruti was looking for a product, with a partner, Fiat was making the 127; other than dating back to 1971, it was a two-door car.

The IMRB survey and the limited success of the Standard Herald convinced the Maruti core team that a two-door car would not work in India. Indian families clearly preferred four-door cars. Fiat was yet to unveil the five-door Uno, and it chose not to inform the Maruti team of its future plans, though the Uno was surely under development then.

In fact, during 1981, the choice of small four or five-door cars in Europe was rather limited. Other than the French and the Italian models mentioned above, the only other small cars available from the big international car-makers were the Mini and Austin Metro from British Leyland, the Volkswagen Polo from the German giant, and two models made by the European arms of GM (the Opel Kadett, badged the Vauxhall Chevette in the UK) and the Ford Fiesta. But all of these were two-door (three-door hatchback) cars.

The Mini was cheap, but the basic design dated back to 1959. True, there was a more modern derivative being made in Italy, the Innocenti Mini/de

Tomaso, but it too, had just two (three) doors. The Austin Metro did get a four-door version but that was much later, in 1985.

British Leyland offered the Ambassaador-sized Morris Ital's complete set of toolings, which in some way may have made sense, as Maruti would have fully owned the rights to the car. But the Maruti core team did not take that route, probably fortunately, as the Ital was just a face-lifted version of the Morris Marina, a car launched by British Leyland way back in 1971. Though the team had been misled into believing the Ital was a recent model launched in 1980, they had read the negative press the car received.

Finally it was Volkswagen that was the most enthusiastic of all the European car-makers. Right at the start it offered the Gol, a small car it had developed specifically for the Latin American market, launching it in 1980. It was cheap and featured the VW Beetle's flat four-cylinder engine, but located at the front, driving the front wheels. But the one problem was the Gol too was just a two (three)-door hatchback.

Volkswagen also offered the Polo, the Derby, the Golf and the Jetta, all cars it had been making in Germany. The Polo and its saloon version, the Derby, had the same shortcomings as the Gol, they were both two-door cars, though the Polo was a three-door hatchback. The legendary Golf had a variant with four doors, as did the Jetta but both these cars were fairly big, (the Jetta was close to the Ambassador in size), and would have cost as much to make, if not more.

To tackle the issue of foreign exchange, Volkswagen also offered to set up an engine manufacturing plant, from which it proposed to buy back powertrains. This offer seemed to be the best and Maruti had almost narrowed down to the German car-maker when it got a message that Volkwagen was not interested anymore; the then Chairman, Toni Schmucker, shot the idea down, believing that investing in India was a waste of time and resources.

By late 1981, it was back to square one for Maruti Udyog and its core team, in more ways than one. By the end of the year, Moolgaokar quit the Board of Maruti Udyog over differences about the direction the company ought to be taking. Moolgaokar believed that it was more viable for Maruti to go into the business of making light commercial vehicles and that there was no future in car manufacturing. So much for experience and foresight…

It was at that point that Maruti looked at Japan and its automobile industry as the only alternative left! Krishnamurthy, Bhargava and Gupta's visit to the Tokyo Motor Show of 1981, was an eye opener. Their information had previously come mainly from *The Daily Express World Car Guide*, which had its limitations as it focused mainly on the European car industry, with scant coverage of the somewhat insular Japanese auto industry. The Frankfurt Motor Show was mainly a European car show.

By 1981, Japan was the world's largest car manufacturing nation, with as many as nine thriving and independent car-makers competing not just in Japan, but across global markets ranging from the USA to Zambia. With a total vehicle production of over 11 million automobiles in 1980 (of which cars accounted for about seven million), Japanese automobile production represented 28.5 per cent of the world auto production of 38.6 million that year, significantly ahead of the US (over eight million vehicles, representing 21 per cent of the world market), Germany (10 per cent of the market with 3.9 million produced), and France (9.7 per cent with 3.4 million).

Yet the Japanese car industry was decades younger than the other three. The first automobiles made in Japan dated back to before WWI, manufactured with European collaborations; the most notable being the Mitsubishi Model A from 1917, a car based on the Fiat Tipo 3. Among the other brands that became major players in Japan later, Isuzu began by collaborating with Wolseley Motors (starting 1918); Nissan's first cars were developed from the Austin 7 and the Graham-Paiges in the 1930s; and the early Toyotas were inspired by the Chrysler Airflow. Interestingly, one of the fathers of the Japanese car industry was the American engineer, William Gorham, who established a three-wheeler manufacturing venture called Jitsuyo Jidosha Seizo Company, which, in 1926, merged with the Kwaishinsha Automobile Manufacturing to create DAT Automobile which later evolved into Nissan Motors, launching the Datsun brand name.

Just about predating India, Ford set up operations in Japan, in 1925; GM in 1927; by 1936, the two had assembled a combined total of 208,967 vehicles in the Land of the Rising Sun. In the same period, the fledgling Japanese car industry had made just 12,127 automobiles. In 1936, the Japanese Government passed the Automobile Manufacturing Industry Law to promote a local industry and by 1939, the foreign car-makers were out of

Japan. The local industry concentrated mainly on truck manufacturing for the war effort. After WWII, a much bombed and battered Japan needed to look anew at the automobile industry.

Given the national emphasis on post-war reconstruction, the initial concentration was on providing the cheapest form of personal transportation possible, other than public transport. Thus, trucks and buses were seen as a priority sector. The Government also decided to encourage the motorized two-wheeler industry. Not unlike India, the Japanese Government in the post-war period, believed that certain industries needed to be controlled to better manage limited resources. The car industry came in for control, whereas the two-wheeler industry did not. A free-for-all ensued, with 50-odd companies plunging into the motorized two-wheeler business; amongst them Honda Motorcycle Company (starting 1949), Suzuki Loom Works (1952) and Yamaha Motor Company (1955).

After the bloodbath that followed, just four of the 50-plus companies survived, with Honda, Yamaha and Suzuki, along with relative latecomer Kawasaki (1962), becoming the Big Four of the motorized two-wheeler world, decimating the world's biggest motorcycle manufacturing nations (including the UK, the leader till the 1960s), and reducing the Italian, American and German bike industries to two-bit side acts.

The Japanese car industry was less cut-throat but everyone who jumped in did not survive. Brands like Prince, Hino, Ohta and others, disappeared or were absorbed by rivals. Yet, by the end of the 1960s, as many as nine had survived the baptism by fire — the same nine the Maruti core team came across at the Tokyo Motor Show of 1981. Of these, Mitsubishi, Isuzu, Toyota and Nissan, were car-makers dating back to the pre-WWII years. Mazda did start off with a three-wheeler in 1931, but its first real (four-wheel) car was made in 1960. Subaru, Suzuki, Daihatsu and Honda were relative industry newbies, born as recently as 1954, 1955, 1957 and 1963 respectively; the oldest just 27-years-old, the youngest barely 18, when the Maruti trio arrived at the Motor Show in 1981.

Unlike the automobile industries of the US, Germany, France, Italy and the UK, the Japanese industry had no one iconic people's car that acted as an engine of growth, propelling an entire industry. Japan did not have a Ford

Model T, a Volkswagen Beetle, a Citroën 2CV, a Fiat 500 or an Austin 7. It had several such exemplars. And that had something to do with legislation and Government regulation. To encourage the production of small, fuel-efficient vehicles that would be of considerable use for small businesses and shopkeepers, the Japanese Government introduced a system of favourable taxation and lower insurance for 'kei' or small cars. Accessible to people who could afford to buy motorcycles but not necessarily full-sized cars, the kei cars started gaining in popularity, simultaneously offering the newer entrants into the car business an easier entry into the industry.

For successful motorcycle manufacturers like Suzuki and Honda, it was a natural move from making bikes to small cars which could use the bike engines they were already manufacturing. For Mazda too, it was a logical move from three-wheelers to cars, as it was for Daihatsu (known before 1951 as Hatsudoki Seizo). For Subaru's parent company, Fuji Heavy Industries, it was a move away from the aircraft industry in which it had heavily invested during WWII, to car manufacturing.

Initially, the engine size allowed by the Japanese Government was too small and constrained for acceptable performance, so very few cars got made or sold. In 1955, the legislation was changed to allow increased engine size (upto 360cc) and modified dimensions (restricted to a maximum of 3 metres length and 1.3 metres width). That is when Suzuki entered the fray, followed by Subaru, Daihatsu, Mazda, Mitsubishi, and finally Honda. In 1976, the limiting specifications for kei cars were again changed to 550cc maximum engine size, mandatory emission control, and restricted overall length to 3.2 metres and 1.4 metres width.

So when Krishnamurthy, Bhargava and Gupta went a-visiting Tokyo, they saw a whole host of these tiny little cars — all about 550cc and just short of 3.2 metres in length — cars that were quite a bit smaller than what they had seen in Europe, and quite a bit cheaper too, at about Rs 30,000. Though the team initially met most of the Japanese car-makers, the choice quickly narrowed down to the smaller cars from Daihatsu (which, by then, was a part of the Toyota Group), namely the Mira Cuore, a kei car; the Charade, a size bigger; the Minica kei car from Mitsubishi; the Nissan Sunny, which was a decidedly bigger car; and the Rex kei car from Subaru (which was then 20 per cent owned by Nissan). Honda (which did make a kei car then), was not interested and neither were Toyota and Isuzu; the reaction from Suzuki was lukewarm at best.

The Mazda R360 Coupé, derived from Mazda's first car, the R360. © Gautam Sen

The N360, Honda's first kei car. Courtesy Honda Motors

For all the intrigue, confusion and misunderstandings, one must read Bhargava's account of the complicated negotiations, the ups and downs and the two-timing of Daihatsu, in his book, *The Maruti Story*. A telex meant for Escorts (the tractor and motorcycle manufacturer that was keen to embark on a car project), accidentally came through to Maruti; on reading it became obvious that Daihatsu did not take the Maruti project seriously.

When all seemed to have failed and suitors had turned betrayers, a knight in white armour arrived in the form of erstwhile recalcitrant Suzuki. A senior director with Suzuki Motor Corporation, who happened to be in India to sign up with TVS for a joint venture to make motorcycles, read an article on the Maruti project on a flight, and decided to swing by Delhi to meet the trio from Maruti. For Suzuki, a joint venture with Maruti Udyog seemed like a great opportunity. As it turned out, Maruti became for Suzuki Motor Corporation, the most important deal in its history as a car-maker — a rather unremarkable history till then. At the time Maruti was in discussion with Suzuki, the latter was one of the smaller and less adventurous car-makers in Japan, mainly known for its motorcycles. Though modestly successful in the kei car segment at home, Suzuki had made few inroads into international markets; its product range was limited to several kei cars.

An early bird in the kei car market, Suzuki's first car, the Suzulight, was launched in 1955. It was an advanced car, featuring a transversely mounted front engine driving the front wheels, fully independent suspension front and rear, and a rack-and-pinion steering system — features that were decades ahead of the competition. However, the Suzulight was a rip-off of a German micro-car, the Lloyd 400, down to even the styling; scaled down to fit the kei car restrictions in terms of dimension and engine size.

Ironically, Lloyd disappeared by 1963, whereas Suzuki marched on in a protected market, replacing its first generation car (and its several face-lifted iterations), with the Suzuki Fronte 360 in May 1967. The Fronte, despite its name, was a complete reversal in thinking — the two-stroke 360cc engine was located at the rear, driving the rear wheels. Why exactly Suzuki took such a retrograde step is not known.

In April 1970, Suzuki launched a kei-sized jeep, the Jimny, which was a back-to-basics, four-wheel-drive vehicle. In 1976, with emission becoming an issue, the allowable engine size was increased to 550cc so that car-makers could develop four-stroke engines that would better meet the norms and yet be as powerful as the two-strokers that they were expected to replace. Suzuki was caught on the wrong foot, as all their engines were polluting two-strokers. The changeover to four-stroke engines took a while and took its toll on the company's profitability.

It was finally in 1979, with the fourth generation iteration of its kei car, that Suzuki was somewhat back in alignment with global trends in terms of technology and mechanical configuration. The fourth generation Fronte switched back to the engine located at the front, with front-wheel-drive. Unfortunately, the new Fronte stayed with a rather antediluvian rigid rear axle with leaf springs that compromised both ride comfort and handling dynamics. But the then technology-starved Indians would not know any better, would they?

Both two and four-door hatchbacks were on offer — the two-door version had a real hatchback tailgate that lifted up, whereas the four-door iteration made do with a lifting rear glass for access to the tiny boot area. Initially, the Fronte was sold with a choice of a two-stroke 539cc three-cylinder engine and a 543cc three-cylinder four-stroker. The latter engine was increased in size to 796cc for the export version, and this version, the SS80 as it was codenamed, in four-door form, was the model chosen for India.

The basis for choosing Suzuki and the SS80 combined commercial reasons regarding terms of the joint venture and pricing. Sought almost as a last resort, Suzuki was one of the few car-makers that seemed to take the Maruti project seriously. It was willing to take a 26 per cent stake (not very willingly, one may add, as the company insisted the investment should not exceed US$ 25 million; the option of taking up to 40 per cent was categorically refused). Most importantly, the CKD (completely knocked down) price of the Suzuki car was the cheapest at the equivalent of Rs 30,000, which meant the target price of Rs 50,000 could just about be met if import duties and so on could be tweaked markedly. This was not an issue as the Government was ready to do just about anything necessary, short of directly subsiding, to ensure the Maruti project worked.

With import duties and excise rates reduced significantly, what India was going to get was a car that was markedly cheaper than the choices available then — the Rs 80,000-plus Hindustan Ambassador and Premier Padmini, and the Sipani Dolphin, a Suzuki SS80-sized fiberglass-bodied two-door car that fledgling car-maker Sipani Automobiles launched in 1982, a car that cost around Rs 70,000 then, some 40 per cent more than the Suzuki's target price of Rs 50,000-odd. Around the time Maruti was nationalised in 1981, R K Sipani's three-wheeler manufacturing company, Sunrise Automotive Industries Limited, was given a licence to make cars too, probably as a sop to criticism in connection with the Government's apparent bias towards Maruti. Thus, Sipani was making cars under licence from specialist British car-maker Reliant Motor Company.

Also, what India was getting was a strikingly more modern car. But that was easy, given the Ambassador dated back to 1956, the Padmini to 1962 and the Dolphin (née Reliant Kitten) design to 1975 and its powertrain to the pre-War Austin 7.

The Sipani Dolphin, in estate form. © Gautam Sen

But was the Suzuki SS80 the best choice then for India's first 'people's car' and India's national car project? Given that 'beggars can't be choosers', the choice was limited to the Subaru Rex and Mitsubishi Minica, other than Suzuki, amongst the Japanese kei cars, the Mini, the Peugeot 205, and the Renault 9. According to *Business India*, 12-25 April 1982, Renault also offered 'a future car model for the exclusive use of their Indian partner', a point confirmed by Krishnamurthy. That may have been the Renault Super 5. But it was too expensive, Krishnamurthy points out.

Between the three Japanese kei cars, there was not much to choose; the Mini was outdated and impractical, being a two-door car; and the Renault 9 was too expensive. The Peugeot 205 would have been pricier than the Japanese kei cars (though one must remember the yen was artificially low then and went up rapidly over the next few years). But it was a markedly superior car in every other way, whether in packaging, handling, road-holding, build quality, performance, *et al.* Yes, the price would have been higher, but the Peugeot 205, or for that matter, any of the great small cars from Europe such as the Renault Super 5 or Fiat Uno, was a whole lot more car, and much better, than the Suzuki SS80.

R C Bhargava, in his book, argues that one of the factors that went against the Peugeot was that the car would not have been ready for assembly in India till late 1984, following its release in France in 1983. The fact remains that the Suzuki SS80 was not the car that became India's 'people's car', it was just an interim model. India's 'people's car' was the fifth generation Suzuki kei car, the SB308, launched in Japan in September 1984, a year after the Peugeot 205. By the same logic, the Peugeot 104 or Citroën Visa could have been the interim model before bringing in the 205...

Another argument for Japanese-versus-European was that the former had several models and variants, based on the same mechanicals, whereas the European cars were stand-alone models. Suzuki had promised Maruti not just a car but also a van and a pick-up, all three using the same engine and sharing many components, as well as a jeep-type four-wheel-drive vehicle. In reality, the jeep-type vehicle (the Jimny), shared no components whatsoever with the others. Amongst the others, the only thing shared was the engine. Even the gearbox and driveline were different, with component compatibility not even accounting for 20 per cent in value terms. In contrast, a car like the Fiat Uno provided the base for a family of models that included a bigger saloon, a five-door estate and a van, with component compatibility in the region of 70-80 per cent. But to understand such technical subtleties better, an automotive culture was necessary — a significant lacuna in the team driving the project.

The Maruti trio was sold on Suzuki; equally strongly, Suzuki Motor recognised the potential in joining hands with Maruti. Krishnamurthy, Bhargava and Gupta met with Osamu Suzuki in March 1982, and on 2 October 1982, the deal was signed (comprising three agreements: a licence agreement on technology transfer between Maruti Udyog and Suzuki Motor, a joint venture between the Government of India and Suzuki and a third agreement for the sale and supply of completely knocked down, or CKD, components for the next 10 years). The deal was signed with one very important caveat: the first car had to roll out by 1983, at a price of Rs 50,000. A tall order indeed! But with the full support of the Government in terms of special waivers, lower taxes and import duties, the impossible became possible. So, it came to pass that on 14 December 1983, on the 37th birth anniversary of the late Sanjay Gandhi, Indira Gandhi, the then Prime Minister of India, ceremoniously handed over the key of the very first Maruti 800 to a certain Harpal Singh from Delhi.

The first customer for Maruti Udyog, Harpal Singh, received the key to his car from Indira Gandhi; Dr V Krishnamurthy stands behind her. Courtesy Avik Chattopadhyay

Less than a year later, Indira Gandhi was dead, killed by her own guards. Maruti Udyog Limited, the nation's pride, the maker of the first Indian 'people's car', became , over the years the single most important outpost of the Suzuki empire. And Harpal Singh, at the time of writing, still owns and uses that very first Maruti 800.

LITTLE CAR, VERY BIG IMPACT

When Harpal Singh took delivery of his Maruti 800, there were 135,000 others in the queue to take delivery of India's new marvel. Maruti Udyog, before production began, held a series of road shows of the three vehicles it was planning to make: the car, the van from the Suzuki range of models known as the Carry, and the pick-up truck version of the van. Ten of each were brought in and displayed in every important town, city and metropolis in India. Indians went crazy. Thousands thronged to take a close look at these little wonders, these brightly painted sculptures in metal that were so much more modern and smarter than the ancient Premiers and Ambassadors the masses had got bored to the eyeballs with!

Maruti Udyog launched a scheme to take advance bookings of Rs 10,000 for each vehicle, 20 per cent of the planned value of the vehicle. The Indian public went bananas: 135,000 vehicles were booked, with the fledgling car-maker collecting Rs 135 crore, almost two-third of its planned investment of about Rs 200 crore for the entire project. This contributed, in part, to the fact that Maruti Udyog became a profitable car-maker from the start of its operations, from the interest earned on the booking amounts.

Of the 135,000 Marutis booked, very few were for the pick-up version, as it was petrol-driven. The commercial vehicle market, historically, has had a clear preference for diesel-powered vehicles. About 20,000-odd opted for the van version, called Maruti Van initially (the Omni later), despite the fact that the van was markedly more practical than the car. More than four-fifths of the punters preferred the car, despite the identical pricing of Rs 45,000 ex-factory for the Maruti 800 and the Van. The bookings were put through a randomly generated lottery system which ensured fair allocation.

The Maruti Van was launched exactly a year after the Maruti 800, on 14 December 1984. The star act, though, continued to be the latter, even though Maruti Udyog managed to deliver just 852 cars across the major Indian cities till March-end 1984. These cars, initially in either post-office red or plain white, were absolute attention grabbers everywhere. Everyone wanted a Maruti 800. There was a massive premium on the cars (as much as what the car cost). India's new small car was a great leveller, with owners ranging from shopkeepers to movie stars. For a short while in Indian history, a car democratised the yawning gap between the rich and the middle-class, becoming a cult vehicle for all and sundry.

During the next financial year, April 1984 to March 1985, Maruti Udyog was able to crank out 20,356 Maruti 800s and 2,016 Maruti Vans. Yet the history of India's first people's car had not really begun. Harpal Singh and the 52,000 others after him, who acquired the Maruti 800 till mid-1986, did not buy India's first 'people's car'. It was touted as one, true, and many made the mistake of acknowledging it as such, but it was just an interim model, a teaser, a preamble to the real McCoy that followed.

THE REINCARNATION OF MARUTI

With the wisdom of hindsight it can be said that it was a smart move on the part of the Maruti management to move to the latest version before committing to the not inconsiderable investment for an entire car project. The Suzuki SS80 that was launched as the Maruti 800 on 14 December 1983, was already a four-year-old car; its replacement in Japan was going to be launched the year after, in September 1984. Surprisingly, Suzuki Motor had not explained this to the Maruti management when negotiating and deciding on the models and the model mix, in sharp contrast to the offer Peugeot had made for the yet-to-be-launched 205 (and if one were to go by the story in *Business India*, and Krishnamurthy's confirmation, Renault even offered a future model). Suzuki's intention was to sell off the toolings of a car it was about to phase out.

The first generation Maruti 800 – the SS80. Courtesy Kartik Ganesh

In all ignorance and innocence, Maruti Udyog launched the Suzuki SS80, with the first batch of cars coming in semi-knocked-down (SKD) state, the bodies complete and painted, only the powertrain needing to be put in, the glasses and wheels attached and the seats and dashboard mounted within. Cars made almost completely in Japan, for the people of India. And as Maruti Udyog went about localising the Maruti 800, little did they know it would be a wasted effort. Around mid-1984, the Maruti management realised the SS80 was soon going to be replaced by an all-new car. Logically, it made sense to invest in the toolings of the newer car. Much against the wishes of Suzuki Motor and Osamu Suzuki himself, Dr Krishnamurthy fought for and managed to convince both Suzuki (and the Government of India), that Maruti too, should switch over to the new model. "It wasn't just a matter of getting the very latest in terms of technology, but also the very principal of technology transfer and being a fair and just joint venture partner that as soon as Suzuki had a new model, it ought to be offered to us at Maruti Udyog," points out Krishnamurthy, when interviewed for this book. "Yes Suzuki's intention was to sell off a car that they were soon phasing out. And if we had carried on with the manufacturing of the SS80 we would have been accused of have been party to such a deal."

The definitive Maruti 800, the SB308, which became India's first true people's car.
Courtesy Maruti Suzuki

In September 1984, Suzuki launched the fifth generation Suzuki Fronte as a five-door hatchback, named the CA 71 for the Japanese market and the SB308 for export markets (with the more powerful 796cc engine from the earlier generation). Without doubt, the SB308 was more practical than the SS80, with more passenger volume than the preceding model, a higher roofline and easier access to the boot space from the vertically lifting tailgate. Yet, on the mechanical front, the car was almost identical, with the same 796cc engine, mated to a four-speed gearbox, driving the front wheels. The suspension system remained similar — independent MacPherson struts at the front, following most small car practices then, but a rigid axle at the rear, sprung by old-fashioned leaf springs. That was the weak point — a somewhat out-of-date suspension system that gave the Maruti 800 a hard, unforgiving ride while compromising road-holding.

The newer SB308 was no better. The ride was almost as hard as the earlier generation model and handling a little poorer as the new car had a slightly higher centre of gravity. But these were issues that did not matter. After decades of the Premier Padmini and the Hindustan Ambassador, the Maruti 800, with all its shortcomings, was more than a refreshing sea-change over the older set of rattletraps. The Maruti 800 not only became a car that was more accessible to the hoi polloi, but became a status symbol and spelt freedom for the fairer sex, being an automobile that was easy to drive and manage. Small was not puny any more, lightweight was not weak and many preconceived notions of the automobile changed. The Maruti 800 became the harbinger of a new India.

India's love affair with the Maruti 800 continued as Maruti Udyog drew up plans to replace the SS80 with the SB308, which, in Maruti nomenclature, became the MB308. The latter was launched in August 1986, after some 52,000-plus of the first-generation Maruti 800s (SS80 model) had been made, with Maruti Udyog assuring customers whose bookings were still pending, that their priority would not be affected.

Localisation plans also shifted to the second-generation car, creating a situation whereby the first-generation Maruti 800 became an 'orphan' in the supply of spare parts, making the maintenance of a car from the early years an expensive proposition. Soon after the second-generation 800 was introduced, Maruti Udyog decided to call for a second round of bookings, in September 1986. The newer car seemed to have even greater fans — 157,000 people booked the new Maruti 800, setting it on the road to becoming India's first real 'people's car'.

During the first year of the MB308, the financial year April 1986 to March 1987, Maruti sold more than 50,000 units (to be exact 52,344 cars) of the 800, a first for any car in India. Also, for the first time, Maruti Udyog's sales of vehicles totaled more than that of the rest of the passenger car industry's put together. Maruti sold 74,511 of the 800, the Van and the Gypsy four-wheel-drive vehicle (which had been introduced in December 1985), in an industry total of 125,894 cars sold during 1986-87, accounting for almost 60 per cent of the Indian car market, and marking the beginning of total dominance by this public sector car company.

It took Maruti Udyog a little over 11 years to produce a million cars but less than four years to make the next million, by 1998; of these, the 800 comprised a significant proportion. By February 1997, Maruti Udyog had produced its millionth Maruti 800. The two millionth mark was crossed in April 2003. The history of the Maruti 800 was thus also the history of Maruti Udyog the car-maker. Reminisces R.C. Bhargava, "I doubt there was anybody who thought 100,000 cars could be done, let alone a million cars!"

After the famous economic 'liberalisation' of 1991, international car majors began looking at India seriously and PSA Peugeot Citroën was the first to enter into an (ill-fated) joint venture with Premier Automobiles. Daewoo followed, inking a joint venture agreement with DCM, taking over from Toyota, DCM's erstwhile partner in manufacturing light trucks. General Motors tied up with Hindustan Motors and Honda entered into a tie-up with Shriram Industrial Enterprises Ltd.

The Peugeot 309 from the Premier-Peugeot joint venture. Courtesy Premier Ltd

But it was not until 1998-99 that Maruti saw serious competition in the form of the Santro from Hyundai Motors India Limited, the Matiz from Daewoo, and later, the indigenously developed Indica from Tata Motors. But that is another story, another chapter. The competition seemed to have caused a drop in Maruti Udyog's sales from 327,240 to 309,544 for 1997-98, which until then had been going from strength to strength. By 1996-97, Maruti's market share had risen to an impressive 79.6 per cent of India's passenger car market; in the financial year 1997-98, Maruti's market share was a staggering 82.7 per cent. Mostly thanks to the Maruti 800!

And it was not just the Indian marketplace that was lapping up the Maruti 800. In 1987, following repeated criticisms that Maruti Udyog was a severe drain on scarce foreign exchange — which was only getting worse as Maruti was behind schedule on localisation, yet selling more cars than planned — the management decided to make a special effort at exporting cars. The Maruti 800 became the first car to be exported by Maruti Udyog and the first set of 64 cars was shipped out to East European Hungary in 1987.

In the first full year of exports to Hungary, Maruti shipped out 500 800s, which increased to almost 1,500 in 1990-91 and then to 4,400 the next year. The success of the Maruti 800 in Hungary eventually led Suzuki to establish its first European car assembly facility in that country. Hungary was Maruti's first safe, tentative step into Europe. But it gave the management the confidence to get more ambitious. Western Europe was the next target, and France the chosen country.

Maruti got the 800 homologated — that is, tested and certified in France — and even crash tested, a stringent requirement that it met. Résau Chardonnet, Suzuki's dealer for France, became the distributor for Maruti too. At that point, Japanese car imports into the European Union were controlled as there was a quota system. Since Suzuki preferred selling its bigger cars, Maruti, as an Indian-made car, had the small city car space to itself. The first batch of Maruti 800s landed in France by late 1989. Distributor Patrick Chardonnet organised "...a gala launch at a posh hotel in Paris, with elephants in attendance," remembers R C Bhargava.

In fact, in Chardonnet, Maruti found one of Europe's finest distributors. Having built up his reputation as an importer-distributor for Italian marques Autobianchi and Lancia, Patrick's father, André, who had founded the enterprise, dabbled very successfully with motorsport. Having just missed the podium at the Monte Carlo Rally of 1960, where André Chardonnet drove a Citroën DS 19, he went on to create his own rallying stable, Team Chardonnet, which rallied very successfully a series of Autobianchi A 112s. Among the many rallyists who cut their teeth with Team Chardonnet in serious competitions, prominent names like Michèle Mouton, Pierre Pagani, Bruno Saby and Bernard Darniche figure. In fact, Darniche went on to corner 40 victories in 50 races with the legendary 'Chardonnet Lancia Stratos' painted in the famous racing 'bleu de France'. With such a pedigree and a distribution network that had sold everything from inexpensive Italian small cars by Autobianchi (a Fiat subsidiary brand) to fancy upmarket Maseratis, Chardonnet provided a great springboard for Maruti's innings in France.

MARUTI 800

INDIAN UNIQUE

Une indienne à Paris ! Voilà qui est rare, exotique et... abordable. Unique représentante en France de la production indienne (et aussi la moins obsolète), cette puce à cinq portes

L'Inde, ses temples majestueux, son fleuve aux eaux boueuses, ses clichés couleur locale... sa production automobile. Certes modeste à l'échelle mondiale (200 000 voitures par ans) mais de souche assez ancienne puisque certaines marques ont été

de Fiat 124 et même 110
ciété Standard, après a
Triumph, évidemment,
ciennes Rover à moteu
de création plus récente
tites Reliant britanniqu
gouttes. Rien qui soit ca
le consommateur occ
par maints produits mc
Tout change avec la c
de la marque Maruti Uc
Gandhi, le fils d'Indira, a
tion accidentelle. Il sout
produire en quantités s
tites autos de conceptio
Et c'est avec le japonais
ruti trouvera un accord

A very positive feature on the Maruti 800 in l'Automobile Magazine.
Courtesy *l'Automobile Magazine*

Initial media reaction was most positive. France's leading car magazine *l'Automobile*, in a detailed four-page review of the Maruti 800 in an article titled 'Indian Unique', in its March 1990 issue, spoke about the virtue of the car being the only one in its category with four doors (all other rivals were two-door hatchbacks), how practical it was for the city and how spacious it was despite its tiny dimensions. "With its 'Hindu' philosophy of modesty and simplicity here is an honest car that lives up to expectations," said the magazine, though issues such as the car's braking and ride comfort came in for some criticism. Yet that didn't stop French men and women from buying 2,833 Maruti 800s in the car's first year of sale in that country.

The invasion of England, Malta and Italy followed. In a country where the national icon, Fiat, completely dominated the small-car market, Maruti seemed to have no chance at all. But in taking on Suzuki's dealer in Italy, Autoexpo, as the distributor, a really wise choice was made. Romano Artioli, Autoexpo's owner, was a car aficionado who had a collection of old Bugattis. On the suggestion of Ferruccio Lamborghini (of Lamborghini fame) and his former chief engineer Paolo Stanzani, Artioli had just about embarked on a project to re-launch the Bugatti marque in the late 1980s from money that he had made importing and distributing Lotus and Suzuki cars in Italy. Around the time when the Maruti 800 was being launched in Italy, Artioli was also busy launching 'his' car, the Bugatti EB110, which at that point of time was the fastest car in the world. Obviously, Artioli didn't think it was prudent to compare the Maruti 800 to the Bugatti EB110. But that did not stop him from comparing the 800 to a Rolls-Royce!

Autoexpo put out a delightful tongue-in-cheek ad in several of the car magazines that showed a side elevation black-and-white photo of the Maruti 800, with a big and mighty Rolls-Royce Silver Cloud behind it, with the text (translated from the Italian) saying: *Just like a Rolls-Royce, the Maruti 800 too has four doors, leather interiors, an audio system and air-conditioning: what more do you need?*

As in France, the 800 received good press in Italy too. "Let's put it this way. We were curious to evaluate the Maruti 800," said Italy's leading car magazine *Quattroroute*, adding, "Firstly, because of where she comes from; secondly, because we were interested in knowing more about such a small car that features air-conditioning and leather interiors."

Auto Oggi, another leading specialist Italian magazine, undertook a comparison test between the Maruti 800 and its two main rivals in the Italian market, the Italian-made Innocenti Small 900SE and the Spanish Seat Marbella Jeans. On a feature-by-feature scoring system, the Maruti tied with the Innocenti (which, incidentally, was a more modern Italian interpretation of the British Mini), ahead of the Seat (a Spanish derivative of the ageing Fiat Panda), scoring 10 out of 10 in the areas of equipment and amenities, and value for money. Where the Maruti lost out was in its styling, which was thought to be dated, and its mechanicals, which the Italian reviewers thought lacked refinement and smoothness.

Yet, the Italians took to the l'il car and by the time sales came to an end for the Maruti 800 — in 2004 — for India's first 'people's car', the people from the land of Leonarda de Vinci had bought more than 19,000 of this 21st century work of plebeian metal. (Ironically, Artioli's Bugatti production totaled all of 139 cars, before the headline-grabbing car-maker went broke). By 1993 Maruti had exported more than 50,000 of the 800s thanks to the combined enthusiasm for the 800 in countries like France, Italy, Holland and Hungary.

The Maruti Zen was launched as a replacement for the Maruti 800.
Courtesy Maruti Suzuki

THE REINCARNATION OF MARUTI

It was in 1993 that Maruti Udyog launched the Zen, the car that was designed to replace the 800 in the export market, and eventually supplement the 800 as Maruti's mainstay for the domestic market too. With impending emission legislation, the management at Maruti was told by Suzuki Motor that the engine of the 800 couldn't be developed to meet those norms. (That engine, with suitable changes, still meets ever more stringent norms such as the Bharat Stage IV, which is equivalent to Euro4.)

The decision was taken to make a car that would replace the 800 in the export market in the early 1990s. But the Zen never quite sold in the numbers hoped for, till a couple of 'price corrections' improved the situation. The Zen's lack of success in the domestic market also translated into a modest export sale of just 123,840 for the car that was touted at one time as the car that would be king, the car that would replace the 800. In comparison, as many as 237,494 of India's first 'people's car' had found buyers in foreign lands till March 2013. By the time the Zen was phased out (in 2006) the car had sold just 640,160 units in India, and a total of 764,000 including exports — not even a quarter of what the 800 has gone on to do, that of 2.95 million sold in India and abroad, by the time it was phased out, in January 2014 — a product lifespan of 28 years (or 30 years if you take its launch in Japan as the starting point)!

True, the car saw some updates in terms of cosmetics, suspensions and, most importantly, engine. The cosmetic and suspension improvements came in 1997 with the incorporation of a wraparound headlamp and new grille and an isolated trailing link rear suspension (which ameliorated the ride quality marginally), improvements that Suzuki had already executed in Japan in July 1986!

The engine, though, saw several steps in modifications, specifically the replacement of the emission-inefficient carburettor for a fuel injection system and improved engine mapping, along with the changeover to a five-speed gearbox from a four-speeder. All of this contributed to retaining the popularity of the 800, with sales climaxing at over 20,000 units as late as March 2003 — 17 years after the car's launch, proving that there was still life left in the old gal!

Interestingly enough, clones of the Maruti 800 are still in production across the border, in Pakistan, where the Suzuki SB308 is sold as the Mehran. And across another border, in China, the SB308 is still being made by Chinese car-maker Jiangnan and sold as the Zotye JN Auto, since Zotye took over Jiangnan. A version of this was also made in Tunisia for the African markets till 2005 and badged as a Peugeot JN Mini!

More than 2.95 million Maruti 800 cars were made over 28 years.
Courtesy Maruti Suzuki

Suzuki had also sold the licence and the tooling for the SB308 to several Chinese car-makers and clones of the Maruti 800 can be seen in China, chugging along with badging as varied as Chang'an SC 7080, the Jiangbei Alto JJ 7080, or the Jiangnan JNJ 7080 Alto or Xian Alto QCJ 7080.

What was to Japan an interim fifth generation model (with a planned obsolescence of five years), of a line now into its 10th generation, became the people's car not just for India but to an extent, Pakistan and China also. Of course, its impact has been the greatest in India, where the Maruti 800 literally put Indians on an automotive diet that continued unabated as India's first true-blue people's car for some 28 years. Amusingly enough, the 800 was never intended by its designers to be a people's car — it was just another competitively priced kei car meeting a Japanese-specific market requirement. But thanks to the strange twists of fate, it became the car that put the world's second most populated nation on wheels.

Which begets the question: could any other similar car with similar pricing and specifications as well have become India's first 'people's car'? R C Bhargava has his doubts: "It's not just a question of the right product. The right partner and people involved are no less important." Bhargava has always maintained that Osamu Suzuki and the senior management at Suzuki Motors were no less crucial in the success story of Maruti Udyog and the Maruti 800.

Of course, his detractors will point out that the success story of the Maruti 800 and Maruti Udyog, the car-maker, had very little to do with the collaborator or Mr Suzuki, but more to do with the clear mandate from the Government of India that this project had to succeed at any cost. "Every rule of the book that could be bent to make the project work was bent … every rule-bending resorted to," says an industry expert who wishes to remain anonymous.

"To start with, licence was given to just one other insignificant car-maker, Sipani Automobiles," says the expert, "then import duties were reduced drastically, as was excise. Almost complete cars were allowed to be imported initially. Maruti was always behind on its localisation commitments, yet the government turned a blind eye to that. Lower excise for fuel efficient vehicles was just an excuse to justify minimal taxation. And then when Rajiv Gandhi came to power and decided to 'broadband' licensing, that was supposed to allow other four-wheeler manufacturers to make cars too, their joint venture proposals were never given a green signal, so that Maruti could be continuously protected. After all that if you cannot succeed, then you need to be incredibly incompetent!" Krishnamurthy concurs, saying, "Yes, any other similarly priced small car from Japan would have succeeded for sure. Of course, working with Suzuki was easier as versus giants like Mitsubishi, but the initial success of the Maruti 800 and Maruti Udyog is mainly thanks to the Indian management."

Yet, there's no denying that Maruti Udyog was a favoured child and the Maruti 800 had a charmed early life. Maitreya Doshi, managing director of Premier Limited, points out that the government largesse in terms of tax and excise benefits gave Maruti "at the least a 15 per cent cost advantage over the others." But that was so for the first decade of the company's existence. Existing car-makers such as Hindustan Motors, Premier Automobiles and Standard Motors, all had the opportunity to work out new collaboration deals to bring in cars and designs that could have competed with the 800. Yet they chose to bring in outdated designs with outmoded technology, launching cars like the Hindustan Contessa (based on a Vauxhall design from 1972), Premier 118NE (a Fiat design, via Fiat's then Spanish subsidiary Seat, from 1966) and the Standard 2000 (a Rover from 1976, powered by an engine that went back to the 1950s). Of these, the 118NE, at its launch, created quite a buzz, with some 108,000 cars booked and Rs 118 crore garnered by the car-maker, but quality issues saw the waiting list dwindling rapidly.

The launch advert for the Premier 118 NE. Courtesy Premier Ltd

It was only by the mid to late 1990s, after the 'liberalisation' of 1991, that Maruti Udyog had serious competition. Yet, not only did the 800 survive, it thrived, despite a crowd of very worthy competitors launching several fine small cars, from the Fiat Uno to the Tata Indica.

FIAT UNO

The Fiat Uno, the car that could have seriously challenged the Maruti 800.
Courtesy Premier Ltd

When Premier Automobiles tied up with Fiat of Italy to announce the launch of the Fiat Uno, 290,662 punters put down almost Rs 600 crore as booking deposits (creating a record that is still unsurpassed). Everyone believed the days of the Maruti 800 were numbered. But then trade union militant Datta Samant decided to intervene and history took a different turn as Premier Automobiles' two joint ventures — one with Fiat and the other with Peugeot — came apart, allowing the 800 to remain king of the hill.

Tata's Indica, Hyundai's Santro and Daewoo's Matiz also looked like serious threats when they were unveiled in early 1998. But then they priced themselves out of the 800's segment. It wasn't till the 800's intended successor, the Alto, came along (in September 2000) that the 800 finally had competition. But then it took the Alto another five long years to topple the 800, to begin a new chapter in India's 'people's car' history.

Chapter V
A Car for India

Two years after Ratan Tata took over from JRD Tata as Chairman of Tata Sons, and had already assumed the helm of Tata Engineering and Locomotive Company Ltd (TELCO), he made a dramatic proposal at the annual general meeting of the Automobile Component Manufacturers Association (ACMA) in September 1993, in Delhi,. What Tata proposed was that Asian car manufacturers and component makers should join together and develop an Asian people's car. Of course, no one took the idea seriously. In fact, at that point in time, no one took Ratan Tata too seriously…

But Ratan Tata was giving voice to a dream that had been very much a part of TELCO's history and that of the Tata family, a dream to make cars. Ratan's predecessor at Tata Sons and TELCO, Jehangir Ratanji Dadabhoy Tata or JRD as he was known, was a car guy to the core. Though many people believe that JRD Tata's first love was aviation (being a keen pilot and founder of India's first airline — Tata Airlines, which later became Air India), JRD was no less fond of automobiles, acquiring his very first car — a Bugatti Type 35A sportster — when he was just 21 years old. Getting a driving licence and acquiring a car in 1926, even an expensive and exclusive sports car, was easier than obtaining a pilot's licence, which JRD did in 1929. Yet his foray into airlines preceded that of the Tata group's entry into the automobile space. Tata Airlines was created in 1932, whereas TELCO (mainly to manufacture locomotives), was set up in 1945.

Yet, soon after India's independence, JRD Tata, along with some senior managers, embarked on a visit to Germany to meet and discuss with Ferry Porsche, the son of Ferdinand Porsche, the possibility of manufacturing Germany's first 'people's car' for the people of India. It is interesting to note that JRD, in the Volkswagen Beetle, recognised the potential of a car that Ford Motors had rejected as worthless.

A Car for India

But with the enactment of the Industries Act of 1951, which signaled the beginning of India's 'Licence Raj', JRD and TELCO had to abandon their dreams of making a car. Instead, TELCO received a licence to make trucks, which, in the scheme of things and a planned economy, was seen to be significantly more important than making cars (a segment with just three licence-holders). For truck manufacturing, as many as five licences were granted in addition to that for TELCO.

All the licensees inked technical collaboration deals with international truck majors, with TELCO choosing to go with Daimler-Benz of Germany for its Mercedes-Benz L3500 (later called the L311) series. In the case of TELCO, not only did Daimler-Benz provide technology, it also took a minority stake in the Indian venture. Initially, the Tata Mercedes-Benz trucks were assembled from CKD (completely knocked down) kits, but soon TELCO began localising the production of the increasingly popular vehicle, which was seen as very strong, reliable and fuel-efficient, thanks to its state-of-the-art six-cylinder diesel engine.

As truck manufacturing and localisation progressed, Tata did not lose sight of the possibility of venturing into car-making. When, in 1959, the Jha Committee was set-up to look into the possibility of introducing a cheaper 'people's car' for India, TELCO was one of the first to submit a proposal. In fact, it offered to make a car produced by a subsidiary of Daimler-Benz, the Junior, by DKW (see Chapter III for more details). At a proposed ex-factory price of Rs 6,950, the DKW Junior could have been an interesting alternative to the cars available then; whether it would have succeeded is another matter. The Junior, like the very slow-selling Standard Herald, was a two-door car, and clearly Indians have always had a preference for four-door cars. Also, Daimler-Benz sold off DKW to Volkswagen in 1964, and DKWs evolved to become Audi, the marque Volkswagen chose to re-launch as a more up-market brand in 1965. Either way, we can only speculate what twists history could have taken since TELCO never did get the licence to make cars.

When, in the late 1960s, the Government of India mooted the idea of a 'people's car' once again, to accommodate the Prime Minister's son and his ambitions, TELCO was one of many who sent in applications. But, as we know, just two were given letters of intent, and only one was able to embark on a project to make a people's car, and it was not TELCO...

Among the early measures taken by Indira Gandhi's successor, son Rajiv Gandhi, supposedly a more liberal-minded Prime Minister, was to introduce the concept of the 'broadbanding' of licences, in January 1985. Thus, any four-wheeler manufacturer could make any form of vehicle with four wheels — a tractor-maker could make trucks or a truck-maker cars.

TELCO saw an opportunity there, so a technical collaboration agreement was signed with Japan's most progressive car-maker then, Honda, to make the Honda Accord. Several other four-wheeler manufacturers signed collaboration agreements with international majors: among others, motorcycle and tractor manufacturer Escorts with PSA Peugeot Citroën, to make the Citroën 2CV, Premier Automobiles with Nissan for the Sunny, and Hindustan Motors with Isuzu for the Gemini.

In the meantime, the Government had approved the joint ventures for several two-wheeler makers and four budding light-truck ventures — DCM (the cloth mill people) with Toyota, Eicher (the tractor-makers) with Mitsubishi, Allwyn (who made refrigerators) with Nissan, and Swaraj Tractors with Mazda. Of these, three were soon decimated into oblivion by Tata's in-house design, the Tata 407 light truck. But with car licensing, nothing happened. No approvals came through, and the Government kept dragging its feet... presumably to provide continued protection to Maruti.

TELCO decided to take matters into their own hands. Sumant Moolgaokar, the then Managing Director and a great believer in research and development and in-house capabilities, gave the green signal to develop a small pick-up vehicle along the lines of pick-ups in the US and Thailand. For Moolgaokar it was still a rather conservative truck-based-approach to car-making — remember he had been Maruti Udyog's first chairman and stormed off the Board when he disagreed on Maruti's plans to make cars. By the mid-1980s, Moolgaokar must have been smarting from the success of the Maruti project. "Yes, Moolgaokar admitted later that he was wrong, but I always had very high regard for him and selected Moolgaokar as *Business India's* Businessman of The Year in 1988," explains Krishnamurthy.

A Car for India

It was also in 1988, that TELCO unveiled the Tata 206 pick-up, referred to by most as the Tatamobile. Indian consumers did not exactly take to the pick-up but the Tatamobile provided the base for the development of two vehicles that brought Tata closer to the car market. The first was a five-door estate derivative of the pick-up, a 'car' called the Tata Estate. Soon after, a two-door SUV, the Sierra, was launched.

Tata 206 pick-up, known as the Tatamobile. Courtesy Tata Motors

These were based on the same ladder-frame platform, with the same mechanicals, just with some differences in the wheel-base and suspension design, other than the skin panels, and all sold in modest numbers. Quality was an issue. Although the Tata Estate looked like a Mercedes-Benz clone and the Sumo a lot like a Chevrolet Blazer with a Mercedes nose grafted on, the two vehicles were too truck-like in their vehicle dynamics and drivability, other than being significantly heavier, less fuel efficient and more expensive, than most of the competition.

It was with the fourth derivative of the Tatamobile that Tata saw its first 'non-truck' success. The Tata Sumo, a multi-utility vehicle (MUV), based on the same platform as the Tatamobile and unveiled in 1994, was a distinctively better MUV than all the others on sale in India then. The Tata Sumo set the sales charts afire. It became the fastest-selling four-wheeler in Indian history, chalking up sales of over 100,000 units in less than 18 months after launch. Even the Maruti 800 had taken longer to get to this figure. At the time of writing the Sumo is still being made and sold, with several facelifts and iterations.

The Tata Sumo. Courtesy Tata Motors

The Sumo's quick design gestation (less than two years), from drawing board to production reality, and rapid market success, gave TELCO and its new boss, Ratan Tata, the confidence to embark on a grander vision —to develop India's first India-designed 'people's car'. After Ratan Tata was greeted with scepticism about his 'Asian people's car' proposal at the ACMA meet, the born car enthusiast within him decided to go the independent, in-house way.

Ratan Tata decided to develop India's first India-designed people's car.
Courtesy Tata Motors

A Car for India

Of course, designing and developing a world-class car represented a very different challenge. With the 407 light truck and the Sumo, TELCO had proved it was adept at developing vehicles for the commercial and semi-commercial segments in India. But to design and put into production a passenger car with more stringent requirements and expectations called for very different skill-sets. The lack of success of the Tata Estate and Sierra had already proven the car segment was a decidedly tougher nut to crack. An equally daunting task was to put the car into mass production while maintaining consistent quality. If Tata did manage to succeed, it would change the face of TELCO forever.

As a first step, a product design brief was drawn up. Ratan Tata felt the Maruti 800 was too small, as was the Zen, the car that was supposed to replace the 800. It was a more modern kei car that was an entire generation more up-to-date than the 800. Though the Zen was bigger than the 800, most of the increase in length was ahead of the front wheels to accommodate the bigger engine, as compared to the 'donor' Suzuki model, the Cervo Mode; the passenger space was only marginally more than the 800. Tata and his team decided that a bigger car with significantly greater space for passengers would be more appropriate for the Indian marketplace. Yet pricing and running cost had to remain close to that of the bestselling 800.

Starting work in mid-1994, a core team at TELCO's research and development centre, Engineering Research Centre (ERC), drew up basic specifications and put together a running 'mule' — a rough, lashed-up prototype that showcased the basic configuration of what could be the truck-maker's future small car. For size and specs, Ratan Tata chose to look at the small cars of European car-makers. The Fiat Uno and brand new Punto, the new Ford Fiesta, and the Opel Corsa, were benchmarked.

For the best part of a year, till about the middle of 1995, TELCO worked in-house, at its R&D centre, ERC, on preliminary sketches and iterations, of which Ratan Tata chose two. Three-dimensional modeling of the shape using computers was carried out by the boffins at ERC, and by the end of 1994, TELCO had a running 'mule', giving a rough idea of the overall dimensions of the car and what it would eventually look like. The engineers were looking for a car that would have a flexible motor, small turning circle, easy handling and decent acceleration. Other than running the mule, the engineers at ERC simulated performance in Indian conditions using computer aided design (CAD).

Whilst examining the Fiats, the Ford and the Opel, TELCO's engineers realised that most of the European small cars had both petrol and diesel options, and so they decided to do the same. For the diesel they set a maximum power target of about 55 horsepower and for petrol, around 60, about 50 per cent more than the Maruti 800. But that was necessary as TELCO's small car was going to be bigger and consequently heavier.

By late 1994, Ratan Tata was having some doubts whether the designs being developed in-house were good enough for a car that would hit the road half a decade later, and remain in production for the better part of another decade at the least... Would the design still be up-to-date 15 years later? Starting late 1994, Tata began looking at the possibility of working with one of the European design houses. He made several visits to the UK to check out design consultancies such as International Automotive Design (IAD) and then went to Italy to have a look at famous design houses such as Pininfarina and Italdesign as well. TELCO had already tied up with IAD to provide for design and consultancy on body engineering in connection with the development of the SUV that evolved from the Tatamobile and the Tata Sierra as the Safari, which was unveiled in 1996 and launched eventually in 1998. IAD had already developed the Land Rover Discovery, and Tata saw an advantage in leveraging that experience for the development of India's first full-size SUV.

For the design of the car, Tata believed that Italy, with its tradition of great automotive design, would be the better country to look at. Though the French had led in automotive design till WWII, post-war the Italians took over, as France's great marques like Bugatti, Delage, Delahaye, Hispano-Suiza and others, fell victim to the punitive taxation the French Government imposed on luxury cars; coach-builders and designers all suffered, with the eventual demise of most.

In Italy, Fiat and the other car-makers such as Alfa Romeo, Lancia, Ferrari and Maserati, all supported the tradition of coach-building of specialist cars (many developed on the basis of more mundane city cars), encouraging the art of designing inexpensive, lightweight yet distinctive cars made in limited numbers. This gave rise to several coach-builders and design houses, among them Bertone and Pininfarina, two of the earliest, and the more modern Italdesign, starting as Ital Styling in 1968, when ace designer Giorgetto Giugiaro embarked on his independent design and engineering consultancy. Into the 1990s, Bertone, Pininfarina and Italdesign were the three most important auto design houses not just in Italy but in the world. Ratan Tata went to Turin to check out these three.

At the same time, he decided to look at the new kid on the block, I.De.A Institute. Though founded in 1978 by Franco Mantegazza, it was only in 1988 and thereafter that I.De.A Institute got noticed for its Tipo Quattro project, which preceded the much-hyped platform-sharing concept of Volkswagen's CEO, Ferdinand Piech. I.De.A. developed a range of seven different models not just based on one platform but using the same set of pressings (stampings) for the doors. The Fiat Tipo hatchback, the Fiat Tempra saloon and estate, the Lancia Delta II hatchback, the Lancia Dedra saloon and estate and the Alfa Romeo 155 saloon, were all designed around exactly the same platform and the same set of four doors. Yet the front and rear sections of the cars varied considerably. But except for some industry insiders, most buyers had no idea they were clones.

The Tipo Quattro was essentially the brainchild of the brilliant designer, Ercole Spada, who already had an amazing résumé as head of design for Italian boutique coach-builder Zagato, where he had designed some of the fastest Aston Martins, Alfa Romeos, OSCAs and Lancias, followed by stints at Ford and BMW, where he designed the BMW 5 and 7 Series of the 1980s. Spada took over as Design Director of I.De.A in 1983 and other than working on the Tipo Quattro, also oversaw the design of the Nissan Terrano II and its sibling, the Ford Maverick. He also worked on a design for Bajaj Auto of a monocoque (unibody, with no separate chassis, which is lighter and more modern than one with a body-on-chassis) auto-rickshaw that never saw the light of day. As Rahul Bajaj explained in 1991: "Monocoque or any old coque, what is more important is the cost of production..."

But by the time Ratan Tata went a-visiting I.De.A, Spada had left; in his place as Chief Designer was a young man of Polish origin, Justyn Norek. Norek and I.De.A were then neck deep in another interesting project — designing and engineering Fiat's world car for the developing markets, Project 178, which, other than conceptualising a relatively 'lower-cost' car than the norm in Europe then, also envisaged a sturdy car with robust engineering that would eventually spin-off six different variants, from a three-door hatchback to a five-door estate, and a pick-up too. The five-door hatchback version, the four-door saloon version and the five-door estate versions, in normal and hopped-up semi-off-road trim, would all be eventually made in India by Fiat as the Palio, the Siena, the Siena Weekend and the Adventure.

Whether I.De.A Institute hinted about the project to Tata or whether Tata heard about it on the grapevine, we do not know, but surely the fact that I.De.A was working on a car with specs similar to what Tata had in mind for India, may have played a role as to why I.De.A landed the plump job of working on the design of Tata's first true passenger car. There was also the not-so-insignificant fact that I.De.A was markedly cheaper than either Pininfarina or Italdesign… In an interview with *Business India* in 1998, Ratan Tata said: "There was nothing wrong with the effort internally. But it was not necessary to reinvent the wheel. I felt that we should look for the best options available in the world in terms of design of the car."

Thus, in early 1995, the collaboration between TELCO's ERC and I.De.A Institute began. Though Ratan Tata and the TELCO people have always stressed that I.De.A was brought in essentially for the styling, the fact remains that I.De.A were needed for the engineering of the body too. Tata's small car was planned as a proper modern monocoque-bodied car that would not have a chassis, as had been done for the Tata Estate or Sierra, but the skins would provide the structural rigidity. TELCO had no experience in developing a monocoque, whereas I.De.A had built its reputation more on its engineering skills than its styling finesse.

Wth the Tipo Quattro project and the project that preceded it — one for Fiat, nicknamed the VSS, or Veicolo Sperimentale e Sottosistemi, in English, subsystem experimental vehicle prototype — I.De.A had explored a very interesting body engineering idea — that of modularity. As Franco Mantegazza, the late CEO of I.De.A said when the author visited the I.De.A facilities in October 1990: "Our first job was for the Fiat group. We made an experimental car, the VSS, which has never been put into production. But the concept was revolutionary. It consisted of a space-frame cage or structure, and the models differ only by the skin. Sometimes different types of vehicles could have the same doors, same windshield. What would differ would be the volumes. The vehicle could be a two-box hatchback or a three-box saloon, or an estate. This modularity of the car is of important concern to all of us at I.De.A."

What it translated into in terms of engineering was a design that gave major strength and security to the car. Moreover, the concept simplified the production process so that several types of vehicles could be made on the same production line, leading to a drastic reduction of automation investment and hence the cost of the car. As Mantegazza explained: "Indeed, huge investments cannot be performed on a single model. If the model in question proves inadequate to the demand, a company cannot accept the loss. Most manufacturers, for example, spend a significant part of their share capital for the development and production of a new car. They incur a huge risk."

Taking on a design consultancy like I.De.A made a lot sense and involved a lot less risk. As Ratan Tata said, there was no point in reinventing the wheel. I.De.A's brief was quite clear — to design a car that was significantly more spacious than the Maruti 800 without making the overall dimensions much bigger. In fact, in an interview with Sanjay Narayan, editor of *Business World*, Ratan Tata was quoted as saying that TELCO's small car would be "...as big as a Maruti Zen, yet provide the space of an Ambassador, at the price of a Maruti 800".

The Ambassador was legendary for space and comfort the car provided to its passengers, a *raison d'être* why it was still selling in significant numbers in the 1990s and even at the time of writing, in 2013. But how to provide that kind of space in a car the size of a Maruti Zen? And of course, the price benchmark was the bestselling Maruti 800.

FROM DESIGN BOARD TO FACTORY FLOOR

So in early 1995, a co-ordination group made up of six experts from ERC, was established. The mandate was to match the cost of the Maruti 800 while making a markedly more spacious car. In time, the original six became 15, who carried out detailed analysis of the cost of each of the components, given the Maruti 800 price benchmark. By mid-1995, the group had figured out the target cost of each component. The engineering team then got working on the specific designs. In the event of cost overruns, they changed the design or manufacturing technique without sacrificing quality.

At a very early stage, TELCO involved its suppliers. In fact, to address this issue, Ratan Tata also initiated TACO, or Tata AutoComp Systems Ltd, a group of companies that took Tata into the component manufacturing business, in joint ventures with several international automotive component majors. TELCO's truck business had a fairly vertically integrated manufacturing system, with most crucial items being made in-house at TELCO's facilities in Jamshedpur and Pune. With the car project came the realisation that it was necessary to have the involvement of international component majors who would not only bring technology and quality to the equation but also the substantial investments required for many of the more modern mechanical aggregates.

TACO signed up with a prominent French component-maker, Sommer Alibert, for the dashboard, trims and bumpers, and with Toyo Japan for the radiators, setting up Tata Toyo. Tata Johnson Controls was the company created to design and provide seats and the interior trim, whilst Tata Ficosa, a joint venture with Ficosa of Spain, provided the mirrors. Tata Yazaki was another joint venture for the electricals. All the partner-companies were world leaders in their respective fields.

TELCO reportedly did not hesitate to squeeze its suppliers and built up a reputation for being aggressive, keeping costs low yet demanding high quality components. When approaching the component makers, TELCO had clear cost and performance criteria, life and reliability targets. And a lot of the component development was a joint effort between the engineers of the component-maker and the ERC engineers, who were present at the component-maker's facilities throughout, either involved in the development or supervising the same.

For the braking system, there was a clear directive from Ratan Tata to use only non-asbestos brake linings, for environmental reasons, as asbestos linings are harmful to the workers who make them. At that time, more than half the Maruti cars in production had asbestos coatings. Tata was not happy about that and insisted the stringent European standards be met. It led to a situation where the initial 2000 brake-linings were imported from a Japanese manufacturer so as to give Rane Brakes and TVS Sundaram time to develop the indigenous systems to meet the performance target and environmental criteria laid down.

TELCO gave the component-makers a target cost for each item and asked them to amortise tooling costs (quite heavy for many components), over the first 100,000 sets. To upgrade the quality of vendors, a special group for improving the quality of suppliers (SQIG or Suppliers Quality Improvement Group), was formed within TELCO, to perform regular audits of supplier quality standards and to establish stringent standards for manufacturing processes. Most senior executives were in constant contact with the suppliers, regularly meeting them to take decisions on the spot and reduce development time.

A feature of the whole exercise was Ratan Tata's constant participation in all aspects of the design and development of the car. He was available 24 hours and exhibited total commitment from day one. Once I.De.A had been signed, the design process moved to Italy, with the Italian design house providing valuable exposure to the designers and engineers from TELCO. Two designers from ERC were stationed at I.De.A for 20 months and several others were sent for three-month stints. Ratan Tata and other senior executives, visited I.DE.A at least five times, each time going into every detail of the design of the car, such as body trims, interiors, even the door handles. Tata's involvement was so great that even the choice of colours for the car was mostly his.

Given I.De.A's experience with project Tipo Quattro and then Project 178 for Fiat, the team had the packaging of the car down pat. Within a fairly compact overall dimension, I.De.A was able to get a whole lot of volume by making the car quite high and arranging for higher seats, with the H-point (the point at which the hips of the passengers are) at a comfortably higher level, freeing up substantial leg-room. Also, the higher roofline helped to get more than enough headroom for tall passengers.

Once the packaging had been determined, which, in many ways, was not dissimilar to that of Fiat's Project 178, the Palio (from which I.De.A had learnt a lot), the process of styling began. Several renderings were made and many sketches proposed, some of which were evolutions of what I.De.A had already offered to Fiat. The official presentation of the first set of renderings was in October 1995. Eight proposals were made by I.De.A. Tata and his colleagues narrowed it down to three. The exterior design was finally frozen by April 1996, and it was to Ratan Tata's credit that the eventual design chosen remains one of the best ever from I.De.A. The design was not unconventional or path-breaking, but just right, with correct proportions, minimal overhangs, a handsome stance and subtle details. The only criticism perhaps being the relation of the wheel-well and wheels, which gave the car an 'under-tyred' look.

Meanwhile, work had been going on for the design and development of the powertrain. For inspiration, TELCO used Peugeot's famous TU petrol and diesel engine family. Launched in October 1986, Peugeot's TU engine was an evolution of the engine family known as X. The interesting fact was that the basic engine block could be used for both petrol and diesel versions: the petrol version was named TU, the diesel version TUD.

About the time the Tata small car was being developed, TELCO and PSA Peugeot Citroën had some discussions exploring the possibility of jointly developing a three-box saloon version of the Peugeot 309 hatchback for the Indian and other international markets. However, nothing came of it.

Sold in the millions, the TU and TUD engines were a good starting point. The initial engine designs were developed in-house at TELCO and the basic engine specifications were frozen by the end of 1995. Then TELCO turned to French engine specialist, Le Moteur Moderne (LMM), after the engine configuration had been finalised. LMM, which had been involved in the development of Peugeot's TU3 version of the TU engine family, got into the act with TELCO to improve engine refinement and make it smoother; reworking and improving the design of the combustion chamber, the cylinder head, and so on. Its greatest contribution was in the design of the fuel injection system for the petrol engine, though TELCO had also developed a simpler, cheaper, but less efficient version with a carburettor. The design of the engine was largely completed by April 1997. The transmission to go with the engine was designed at the same time.

The first set of prototype engines and transmissions were shipped to I.De.A's facilities in Turin in April 1997 and the first two prototypes were assembled by June. The occasion was celebrated with toasting and spraying of champagne. The next important step was taken in September 1997, when the first prototype was assembled in Pune and subjected to extensive testing on TELCO's torture track. A new crash test centre, the first of its kind in India, was built to test both the Tata Safari and the small car. In this test, the vehicle, with a dummy 'driving' it, was crashed against a concrete barrier at a specified speed. Rapidly, the prototype totaled over 100,000kms of on-road testing.

TELCO also built an anechoic chamber for the first time in India, to test the noise and sound issues. An acoustically designed room, insulated against outside noise, the anechoic chamber provides the setting for actually and accurately measuring the decibel level of engine noise, or the level of sound within the car. With extensive development on rubber and window sealing, TELCO was able to match international standards for noise levels within the car, which was around 64 to 66 decibels for the petrol and diesel versions respectively, when the car was travelling at 60km/h.

By the end of 1997, over 400 engineers and designers from ERC were on the project, apart from hundreds of others working on getting the manufacturing facilities ready. For most of the machinery going into the facilities for Tata's car, TELCO's Auto Projects Division was entrusted with the sourcing and purchasing of equipment, from some of the best manufacturers in the world: machining equipment came from Heller in Germany and Ingersoll from the US, Gleason, another American machine tools specialist, supplied the gear-cutting machines, grinders came from SAIMP, Italy, and Buderus in Germany. The high-speed press-line came from German company Erfurt. Also, 70 robots for the seam and spot welding of the bodies were brought from Japan. The very important step of making the dies for the body, went to Japanese specialist, Miyazo.

But the basic plant itself was not new. TELCO acquired a not-so-old Nissan plant from Australia, which the latter had 'abandoned' as part of its cost trimming exercise during its deep financial troubles in the mid-1990s. Durr of Germany was given the task of commissioning and upgrading the plant and the Nissan paint shop TELCO had acquired. Buying the Nissan plant for just Rs 100 crore saved TELCO considerable sums of money and time; with development costs touching barely Rs 206 crore (including Rs 60 crore for I.De.A's design and body engineering), TELCO's car project cost barely Rs 1,700 crore — a fraction of what international majors incur for an all-new car. Though the sum was substantial for TELCO, the equivalent of US$ 400 million, it was peanuts compared to the US$1.5+ billion typically spent by international car majors to develop and tool up for a new model.

So TELCO's first major car project had eventually almost 700 engineers, taking 31 months to complete a project that involved the design and development of 3,885 components, as many as 740 dies and over 4,000 fixtures, specially made for the manufacturing of the car, accounting for Rs 74 crore in tooling and Rs 1,420 crore in setting up the plant. In the process, TELCO's capital cost per car workeed out to Rs 113,000 versus Rs 114,000 for the Maruti Zen, which was in 1997 only a four-year-old car, but markedly smaller than what TELCO was about to unveil in January 1998.

The Tata Indica at Pragati Maidan, January 1998. Courtesy Tata Motors

A Car for India

15 January 1998, was a red letter day in the history of TELCO and India's automotive history. Several significant new cars were launched at New Delhi's Pragati Maidan expo grounds, including India's first two-seater sports convertible and coupé. But the car that really grabbed the headlines was the Tata Indica, India's first home-grown 'people's car'. Even though everyone knew the new Tata car 'that would be as big as a Maruti Zen, yet provide the space of an Ambassador, at a price of a Maruti 800', was going to be called the Indica (*India*+*Car*), they still had not seen the car. So the unveiling was a much-awaited moment. Expectations ran high, with thousands of journalists, politicians and spectators crowding the huge Tata stand at Hall 11. The car did not disappoint. It delivered on the promise of space and size…in a good-looking package.

Miss World Diana Hayden and the Tata Indica at the Geneva motor show, 1998. Courtesy Tata Motors

A Car for India

The Indica was launched in the Indian marketplace by the end of 1998, as Tata had promised. The initial bookings (about 115,000), and expectations, were huge for a car (though priced more than the Maruti 800), was still markedly cheaper than the smaller Maruti Zen. Sadly, early teething problems blunted the public enthusiasm about the Indica. But to the credit of Tata Motors (TELCO had been renamed Tata Motors), it did realise the shortcomings of the Indica and began addressing them.

Dr V Sumantran, formerly of General Motors, joined Tata Motors in 2000, as Executive Director of the passenger car business and head of ERC. One of his first tasks was to address the quality issues of the Indica. Ratan Tata was also keen the Indica be exported, specifically to the developed markets of Europe and other parts of the world.

As part of the plan, Sumantran was tasked with discussing with international majors whether they would distribute the Indica through their networks. One of the first car-makers he spoke to was Chrysler, which did not have a small car in its range. But evaluations showed the car would not have met future American safety norms which were getting more stringent.

At the same time, development work was on for the three-box saloon version of the Indica — the Indigo, to be launched in 2002. As commonality of parts between the Indica and Indigo was an important part of the planning, it was decided that the front firewall (the metal panel separating the engine compartment from the passenger area) and the front longerons (part of the chassis stiffeners), would be redesigned to improve crash safety.

Thus, with these changes and modification to several details to address recurring problems with the front suspension and carburettors and other components, the Indica was re-launched as the Indica V2 (version two). In this version, the car became a much better product. And Tata Motors was able to convince ailing British car-maker MG Rover, to sell the car badged as a Rover in the UK. Though the Rover CityRover (as it was called), came in for a lot of hammering from the caustic British motoring press, specifically BBC TopGear, which rubbished the car, the CityRover is still listed in the top 100 most reliable cars in the UK's reliability-index, even eight years after it went off the catalogue.

A Car for India

Coming back to India and the Indica, the car became the basic platform for Tata's passenger car range. Four years after the unveiling of the Indica, at the Auto Expo of 2002, Tata Motors launched the Indigo, a three-box saloon derivative based on the Indica, and two years later, the Indigo Marina, an estate derived out of the Indigo. In the financial year April 2006-March 2007, Tata Motors managed to sell as many as 144,690 Indicas.

At the same time, at subsequent Geneva motor shows, starting with the 2000 edition, Tata unveiled a series of Indica-based concept cars: the Aria, the Aria Coupé and the Indiva. As a concept car, the two-seater Aria convertible was nothing new — two-seater sports cabrios have been around since the early years of motoring. But what Tata was trying to establish was the potential for such a derivative based on the Indica platform and better still, establish a design language for future Tata models, specifically the face. The smiling grille of the Indica was retained and with headlamps as crescents defining the edges of the bonnet-grille line, the Aria was a very pretty looking two-seater that embodied the new Tata look.

But if the Aria had been made, there could logically have been a coupé derivative too. This possibility was explored with the Aria Coupé (also a non-starter), the year after, at Geneva. More interesting was the third concept Tata showcased at Geneva in 2002: the Indiva — a MPV or minivan concept with much greater potential. Theoretically based on the Indica platform, the Indiva (*Indi*ca meets *va*n) was to use a stretched version of the floorpan to allow for greater volume and was good enough for a seven-seater.

But these concepts remained just that: concepts. Instead, Tata concentrated on bread-and-butter matters. The first generation Indica family, comprising the very popular Indigo and the more modest-selling Indigo Marina, was ageing, so the time was right for Tata to get into the next generation of cars.

In January 2008, at the Auto Expo, where an even smaller car was the cynosure of all eyes, Tata also unveiled its brand new Indica Vista, marking the beginning of its second generation. At the same time, Tata launched the last iteration of the first generation Indica — the Indigo CS. The CS was

Tata Aria sports car concept, pretty but eventually pointless (courtesy Tata Motors)

Tata Aria Coupé, derived from the convertible roadster concept (courtesy Tata Motors)

Tata Indiva concept, an idea with much potential, but never developed.
Courtesy Tata Motors

an Indigo with a shortened boot. It brought the CS into the four-metre tax break category, making it significantly cheaper than the Indigo. Launched in late 2008, the Vista, though similar in concept to the Indica, was all new, significantly bigger than the Indica, with completely new body panels, decidedly better made and more sophisticated. Part of Dr V Sumantran's three-platform strategy for the Tata Motors cars business, the Vista was the first of the all-new X1 platform. And then, in 2009, the next-gen version of the Indigo, derived logically from the Vista, the Manza, followed. The Manza was, like the Vista, based on the new X1 platform and had nothing really to do with the Indigo either. Yet the branding — Indigo Manza — gave the impression it was nothing more than a refreshed Indigo. The Vista too, suffered as most consumers saw the car as a refreshed Indica.

So, what started as India's first all-Indian car in 1998, had become by end 2009, a four-member family spanning not only a wide price range, but had sold well over a million cars. In the process, it had given millions of Indians a very Indian set of wheels.

The all-new Tata Vista had very little to do with the Indica. Courtesy Tata Motors

Chapter VI
People's Car — the Korean Answer

The tata Indica was not the only star at New Delhi's Pragati Maidan expo grounds in January 1998. Other than India's first two-seater sports convertible, the San Storm, two other li'l cars had caught the fancy of the hundreds of thousands of visitors to India's fourth motor show, Auto Expo 1998. One was a cute little car called the d'Arts, the other, also another small car, the Santro, which was being shown after being unveiled in India at the Delhi International Motor Show a year earlier, in January 1997. Both these cars were presented by two South Korean car-makers. The Indian public was just getting to know one of the two manufacturers, as the large and rather expensive Cielo, assembled by Daewoo at a brand new factory near Delhi, had started selling in modest numbers across the bigger cities.

Yet Daewoo or Hyundai (the other Korean car-maker that had unveiled the Santro), were far from being household names in a country that had grown up on a diet of European and American cars, with a smattering of Japanese brands such as Toyota, Honda, Nissan and Mitsubishi — most of which had been brought in by consulates and returning expatriates. Yet, these two Korean car-makers seemed to have ambitions way beyond the European or Japanese car-makers setting up fledgling Indian operations. French giant Peugeot had tied up with Premier Automobiles to make the not inexpensive Peugeot 309; Ford was assembling the rather expensive Escort with Mahindra & Mahindra; GM was making the very expensive Astra, launched by its German subsidiary Opel; Honda was planning to make the not inexpensive mid-sized car, the City; Hindustan was planning to assemble the rather expensive Mitsubishi Lancer; and Toyota was bringing in an unappetising multi-purpose vehicle, the Qualis. None was really aiming for the main segment in the marketplace, where all the action seemed to be — that of small inexpensive hatchbacks, potential 'people's cars'.

The San Storm sports convertible, also unveiled at the 1998 edition of the Auto Expo.
© Gerard Godfroy

Except for Fiat, which, with Premier Automobiles, had unveiled the Uno two years earlier, no one else seemed to have any plans to take on redoubtable market leader Maruti Udyog. But with Fiat and Premier floundering with labour problems at their factory in Mumbai, along came Tata Motors with the Indica. And matching Tata and Maruti spec for spec, price for price, came two unknown Korean upstarts, Daewoo and Hyundai, with their two new small cars.

Korea's history of car-making, till 1998, had been rather short since the industry was barely three decades old. But it is a story that must be told. Not unlike India, Korea had no automobile industry before WWII, as Korea was a colony of the Japanese. It was only after WWII and the partition of Korea into totalitarian North Korea and democratic South Korea, that the first seeds of an automobile assembly-based market were sowed. Not unlike India, South Korea too, had a planned economy; to foster a domestic automobile industry of sorts, the government enacted the Automobile Industry Protection Law in 1962 — something not dissimilar to India's Industries Act of 1951, which saw the start of 'Licence Raj'. The law had three key provisions: assemblers were given tax exemption, parts and components were exempted from import tariff and the import of complete cars was prohibited, like in India, by the late 1950s.

Interestingly enough, this law came through thanks to the intense 'political lobbying by Kim Jong-pil, then director of the newly created Korean CIA, on behalf of Nissan Motors, which along with its local partner Saenara Motors (founded by Park No-jung, a Korean resident of Japan), sought access to the nascent Korean market', according to a study of the Korean automobile industry by Andrew E Green in 1992. So there too, there was an interesting parallel with India...

So, from 1962, assembly line production began in South Korea. Several assemblers were given licences. But localisation was incentivised and local content rose from less than 20 per cent in the mid-1960s to over 90 per cent by the end-1970s. Not unlike India, but still very much behind it, not just in terms of localisation but also production, as all cars made by assemblers in Korea in 1974, totaled just 9,069. Whereas in India, in 1970, the total cars produced — between Hindustan, Premier and Standard — was over 35,000. By 1973, the figure had gone up to an impressive 94,869 in India and though the figure for 1974 is not available, it was more than in 1975, when production had 'fallen' to 74,800 (according to SIAM, the Society of Indian Automobile Manufacturers). By 1980 though, the figure had come down to just 30,989. In South Korea, meanwhile, the production of cars had gone up to 112,314 in 1979, but in 1980, thanks to the second oil shock, demand plummeted to less than half that of the preceding year.

To put it in a different perspective: India produced 76,409 motor vehicles (cars and commercial vehicles), in 1970, as compared to 28,819 produced in South Korea. By 1980, South Korea had made 123,135 vehicles against India's 113,917. Yes, South Korea had overtaken India.

But it was in the 1980s that the South Korean car industry exploded, from rolling out about 55,000 cars to manufacturing close to a million by 1990 (986,751 to be exact) – an absolutely mindboggling growth of 20 times in ten years! In terms of motor vehicles (cars and trucks) produced, India had made 362,655 in 1990 and South Korea 1.32 million.... By 1995, India's production had gone up to 636,000 motor vehicles and South Korea's to 2.5 million. By 2000, the gap had widened — a tad over 800,000 made in India versus 3.1 million in South Korea, making it the fifth biggest vehicle (cars and commercial vehicles), manufacturing nation after Japan, US, Germany and France.

People's Car — the Korean Answer

The meteoric rise of South Korea's automobile industry was a combination of Government favouritism and the ability of a small number of ambitious entrepreneurs to perform, and on occasion, take orders. Those who were able to meet the goals set by Government policy-makers were rewarded with whatever was necessary to ensure continued growth.

Though the very first South Korean 'cars' were lashed-up Jeeps (remnants of WWII and the Korean war), made by the Korean entrepreneur, Choi Mu-seong, followed by Sinjin Automobiles assembling the Toyota Publica, the first three proper assembly operations began from 1962, following the Automobile Industry Protection Law: Kyeongseong Precision Industry, Ha Dong-hwan Automobile Industry Co and Saenara Automobile, the company on whose behalf Korea's CIA chief had lobbied for the new law.

Kyeongseong began by assembling Mazda cars under licence and in time changed its brand name to a simpler 'Kia'. Ha Dong-hwan in time became SsangYong, and Saenara, which was assembling Nissan cars, was taken over by Sinjin, which was assembling Toyotas till 1972. After Toyota withdrew in 1972, Sinjin began collaborating with GM, making GM cars under licence and formed General Motors Korea. But the company then changed its brand name to Saehan, which in turn was taken over by the Daewoo Group in 1978, and the company was renamed Daewoo Motor Co Ltd.

These first three assemblers were soon joined by two more: Asia Motors Company in 1965, followed by Hyundai Motor Company in 1968, to assemble Ford Cortinas under licence.

Then in 1974, as part of the South Korean Government's broad push into heavy and chemical industries, it promulgated the Long-Term Automobile Production Plan, to stimulate economic growth and an indigenous automobile industry. It invited assemblers to come up with ambitious plans to build a Korean automobile industry. The five main assemblers — Kia, SsangYong, Sinjin, Asia and Hyundai — took up the challenge, but it was Hyundai that set the blueprint for the Korean way.

Hyundai, which means 'modernity' in Korean, was established as an engineering and construction company in 1947, by Chung Ju-yung. Son of peasants, Chung Ju-yong left his family (who lived in a village in what is

North Korea), at the young age of 16, and survived doing odd jobs, including that of a rice trader. After the war and the 'liberation' of Korea, Chung Ju-yung founded Hyundai Engineering & Construction Company in 1947, mainly for repairing and maintaining automobiles. Eventually, Hyundai went into heavy construction, which has remained an essential activity of the corporate group.

After the Korean War of 1953, the Americans, specifically Ford, entrusted Hyundai with the repair of military vehicles and construction of barracks for the US Army. In the 1960s, Chung Ju-yung got close to General Park Chung-hee, the military strongman who became President of South Korea from 1963. Thanks to Ju-yong's proximity to the powers-that-be, Hyundai received the contract to make the motorway linking the capital city of Seoul to the important port town of Busan, in 1968. This, in turn, led to shipbuilding, the building of dams, nuclear power plants and so on.

So when the call came to take Korea into the automotive arena, Chung Ju-yung and Hyundai were more than ready; they already had a brilliant master plan. Chung Ju-yung did not see the need for reinventing the wheel, nor did he see the need to follow the Japanese way of copying car designs from Europe; preferring to build an industry independent of the West. He decided to go to Europe to get the car designed on the basis of tried and tested Japanese mechanicals. One of his first steps was to hire an experienced British automotive executive to run the business. George Turnbull, former Managing Director of Austin Morris at British Leyland, was brought in, along with a team of five engineers. For the powertrain, Hyundai signed up with Mitsubishi for the supply and eventual localisation of the latter's 1.2- and 1.4-litre petrol engines used for the Mitsubishi Lancer model of the 1970s.

For design, Hyundai went to Italian design house Italdesign. Dies and moulds for the body came from Ogihara Mould Company of Japan, the presses from France, and funding from the French bank France Suez and British bank Barclays. However, the British management and engineers left soon afterwards over some disagreements and the engineering was completed by moonlighting Japanese engineers.

The Pony, Hyundai's first car, was unveiled at the Turin motor show in October 1974. It went into production by end-1975. Though the first model in production was the four-door saloon, three-door and five-door hatchbacks and an estate joined the range shortly, with a pick-up version added later. As the domestic market was too small, the South Korean Government had insisted the industry be export-led. Hyundai exported its first batch of six Ponys to Ecuador in 1976, in exchange for bananas...

Soon other countries followed: Argentina, Columbia, Egypt and, by 1978, countries in Europe, mainly Belgium and Netherlands. In 1982, the Pony received a facelift, becoming the Pony II, and began to be exported to the UK, and then Canada, where it became the bestselling car as it was the cheapest car on sale there. Sales to the US started in 1986, with the Pony II sold as the Excel, a four-door saloon, and as a hatchback. Hyundai set a record for the first year of sales in the US, selling as many as 168,883 Excels. But it proved to be a flash in the pan as sales quickly dwindled due to quality and reliability issues. In fact, Hyundai faced much ridicule in the US soon thereafter for making unreliable rattletraps. But that did not discourage the South Korean car-maker.

In 1984, Hyundai launched its second car, the Stellar, which was basically the Ford Cortina (still being assembled by Hyundai), but redesigned by Italdesign into a more modern design, and powered by 1.4- and 1.6-litre Mitsubishi engines. With a larger 2-litre engine, the car was launched in the North American market. The Stellar was eventually replaced by the Sonata in November 1985, with its American launch following a few years later. By 1991, Hyundai expanded its range to the Lantra, which was slotted in between the Pony/Excel and the Sonata. And then Hyundai went on to develop its first coupé, the Scoupe.

But most significantly, it was in 1991 that Hyundai launched the 1.5-litre Alpha, its first in-house developed engine. This came by when Hyundai ran into problems with Mitsubishi. In the extreme cold, the engines would not fire properly; when it was detected the problem was with the Electronic Control Unit (ECU) that 'controls' the engine and that it needed to be reprogrammed, Mitsubishi claimed it had sold the engine and not the ECU. So if the ECU needed re-programming, an additional payment would need to be made — a conflict that Hyundai resolved by deciding to build its own engines and subsequently breaking away from Mitsubishi.

By the mid-1990s Hyundai had the capabilities to design and develop complete cars in-house and it did not need an Italdesign or any other external consultant to help out anymore. Yet, into the 1990s, Hyundai cars were still notorious for poor build quality and lack of attention to detail.

It was only in 1999, when Hyundai introduced its 10 year/10,000 mile vehicle warranty programme in the US that its climb back from the abyss of poor quality began. This warranty was finally proof to American consumers that Hyundai was going to stand behind its vehicles and offer a higher level of protection and better quality for its customers. The same deal, though, was not offered in other parts of the world.

BRINGING 'MODERNITY' TO INDIA

The unsuspecting Indian public had no clue of Hyundai's history, or for that matter, of Daewoo's, when the Santro and the d'Arts were shown to the enthusiastic public at the 1998 edition of the Auto Expo. Nor was it aware that by 1998, Daewoo and Hyundai were, respectively, the 15th and 16th biggest car-makers in the world, having produced 727,000 and 704,000 cars, respectively. But it could see the commitment of the Koreans.

It was Daewoo which had made the first move into India, investing in a huge plant near New Delhi. But when the management insisted the Daewoo Cielo was a 'brand new' design, the first seeds of suspicion were sowed as most motoring journalists knew the Cielo was just a face-lifted Opel Astra from the earlier generation.

With the d'Arts, which was renamed the Matiz at the time of its launch, Daewoo was taking a leaf out of the Hyundai strategy and launching a car that, though it had Suzuki mechanicals under the skin, was an all-new design by Italdesign. Daewoo had become a client of Italdesign, replacing Hyundai, which had moved on to its own in-house design expertise. The Matiz was Daewoo's take on a small car, a 'people's car' for markets other than South Korea, which preferred bigger saloons. As the international market, specifically that of Europe, had become important and the Europeans had a penchant for small cars, Daewoo realised that it needed a small car for Europe, which could also work as a 'people's car' for markets like India.

People's Car — the Korean Answer

The biggest problem with the Matiz was its introductory price — too high for a car that was not much bigger than a Maruti 800 and featured a similar powertrain. People did not want to pay Maruti Zen prices for a car that had an engine that was essentially the same as the 800. In fact, the Matiz was actually a redesigned Suzuki kei car, the Alto, from one generation after the SB308 (the 800), which Daewoo had been making under licence, badged the Daewoo Tico. So what Daewoo did was to update Suzuki's 'people's car' to Daewoo's car for the people.

Daewoo Matiz, one of two Korean 'people's cars'. From the author's archives

Despite the huge factory Daewoo had set up in India, its localisation programme was out of whack too and it could not reduce prices without losing a significant amount of money. Also, other than its convoluted history, Daewoo had grown too rapidly in the 1980s and 1990s, fueled (and over-leveraged), no doubt by its ambitious owner, Kim Woo-jung (who had written a book at the height of his success story, titled *Every Street is Paved with Gold*!); so much so that it was in big trouble thanks to the Asian financial crisis of the late 1990s.

But Hyundai was made of sterner stuff. By the time the South Korean car-maker came to India, it had fairly decent in-house capabilities. It had had the humbling experience of succeeding and almost failing in the world's most competitive market, in North America, and the experience of setting up a plant in a foreign country with great expectations and seeing it fail within four years, when Hyundai had to close down its Canadian assembly operations at Bromont, near Quebec, in 1994.

The opening of the Indian car market proved to be a heaven-sent opportunity for Hyundai. With liberalisation, several Indian business houses had approached Hyundai, and the Korean company's executives, in turn, made several trips and surveys in India, engaging in detailed discussions with potential collaborators. But when they saw that it was possible to come into India with a wholly-owned subsidiary, they chose to do so. Not only was their subsidiary in India 100 per cent, so was their commitment. Hyundai Motor India was incorporated on May 6, 1996. Hyundai was clear, that like its gigantic factory at Ulsan, in South Korea, the factory in India had to be near a major port. Soon, it had decided to locate the plant at Sriperumbudur, near Chennai. The groundbreaking ceremony for the plant began on 10 December 1996. The process began to transplant the recently mothballed Canadian plant of Hyundai Motors to India, lock, stock, barrel, paint and press shops.

By early 1997, they had hired B V R Subbu as Director, Marketing and Sales. He had been a star at TELCO, and Ratan Tata's blue-eyed boy at one time; he would be the main architect overseeing the Hyundai success story in India. The initial plan was to bring in the Hyundai Accent, the mid-sized saloon model, following the template Daewoo had set in India — come in with the more expensive car and then, as localisation progressed, bring in the smaller car later — a top down approach to entering a new car market.

But Subbu didn't agree with that strategy: "It appeared quite clearly that the small car segment should be the starting point and we could graduate later into the higher-price segment." What the market needed was a 'people's car'. Subbu believed it made much better sense to go straight for the jugular, the main market segment, that of small hatchback cars like the Maruti 800 and the Zen, which accounted for three-quarters of the total car market in India then. Hyundai had then, in 1997, launched the Atos, the company's first small city car; to Subbu, it seemed to be the most obvious choice for India. Hyundai's senior management agreed.

But there was one problem with Hyundai's in-house home-grown product: it was incredibly ugly. The car had been put through a customer clinic in India where people had been asked for their opinion. Though most thought the car had many virtues in terms of space and comfort, all disliked the look. Hyundai's first small car, designed in South Korea, was an aesthetic disaster. Obviously inspired by the 'tall boy' design of the very successful Suzuki Wagon R which had become one of Japan's bestselling cars in the 1990s, the Hyundai Atos was a rather gawky-looking tall car with an upright backside that liberated a whole lot of inside space within the rather compact dimensions but gave it the look of a car that was going to keel over at the slightest provocation.

Subbu was convinced the car, as it was, would not sell. The Hyundai management came to the same realisation. But then Hyundai could not go back to the drawing board and redesign the Atos as a completely new car for India. So the best compromise was to rework the rear portion and somewhat ameliorate the design, which, of course, called for new dies and tooling for the rear tailgate, the hatch, that is the rear roof section and the rear fenders, the rear bumper and the rear lamp assemblies. It is to Hyundai's credit that it realised the importance of making these changes and incuring the cost of additional dies and toolings for a product that had only recently been launched.

So, at the Auto Expo of 1998, Hyundai unveiled their small car on 13 January, christening it the Santro instead of the Atos. The Atos moniker, in the meantime, had changed to Atoz for the UK market, as Hyundai's marketing people realised the British public would not give a toss for the Atos; so the car became the Atoz, a car that would take you from point A to Z.... For India, a newer, better looking car deserved a different name, hence the Santro. Hyundai also seemed to be in a rush to get it to the market, having realised the importance of being first mover. On 27 May 1998, the pilot production of the Santro began within a record 17 months of groundbreaking, and four months after it had been shown to the public for the first time. By 23 September 1998, the first batch of Santros was in showrooms across India, beating the other two stars from the Auto Expo to the marketplace.

Priced between the Maruti 800 and Maruti Zen, the Santro did not exactly set the sale charts afire to begin with. The expectations were more modest for a car that looked rather practical, did not have the sheer elegance of the Tata Indica or the lovable cuteness of the Daewoo Matiz. Yet, there was no denying its functional quality, including spacious interiors that seemed to match the markedly bigger Fiat Uno. The 999cc four-cylinder engine was smooth and linear in its power delivery, the engine note refined and unobtrusive, with low-end drive-ability seemingly better than that of the Maruti Zen. Though the body rolled when cornering hard, road-holding seemed to be good enough, and the ride comfort was decent. The air-conditioning was very effective and the top-end version came packed with features such as power steering, electric power windows for the front pair, an audio system, a rear screen wash-and-wipe system, and body-coloured bumpers with fog lamps. Here was a car for the people, but with appropriate baubles and bits attached.

The Hyundai Santro at its unveiling. Courtesy Hyundai Motors

At a price markedly lower than the Zen, the Santro seemed to make good sense, even if it did not spell love at first sight. But then no one at the time had really heard of Hyundai. Did the Koreans make cars? Did they know how to make cars? Were they good enough to compete with the Europeans and Japanese? There were simply too many questions; too many doubts.

That's when the role of marketing and a brilliant advertising campaign came into play, a clear first in the Indian automobile scenario. Hyundai Motor hired Saatchi & Saatchi to think up a campaign that would break into the Indian marketplace. They brought in the concept of a brand ambassador for Hyundai's first car, and not just any brand ambassador but one of Bollywood and India's biggest rising stars — Shahrukh Khan. In a series of television commercials starting April 1998, before the Santro had been launched, Hyundai began to address the many doubts and issues surrounding the brand, the car, the abilities of the Koreans, and the commitment of the car-maker to India.

The first phase of the advertising campaign, from April 1998 to June 1998, had Shahrukh Khan being chased by a Hyundai executive, a certain Mr Kim, trying to convince the Indian actor to endorse the product. The very first ad introduced the brand and the company with a very subtle statement from Kim: We settle only for the best. Then the teaser ads that followed had Kim trying to convince Shahrukh Khan that Hyundai was serious about India — that it already had in place dealers, suppliers, workers, etc — about quality issues and the brand Santro.

In the second phase of the advertising campaign, which ran from July to September 1998, Khan was shown asking the kind of questions a typical Indian consumer would ask: Who or what is Hyundai? How can Santro be good enough for India roads? Will Hyundai be able to match customer requirements and provide proper after-sales services? At the end, Khan pronounced: "I am convinced", and in the process convinced many Indian consumers that the Hyundai Santro could indeed be the "complete family car" they were looking for, a car that had more space and features than the people's car of the period — the Maruti 800.

This very high profile celebrity association with some innovative media buying and aggressive press coverage, created enough hype for the Santro and sales took off, contrary to expectations. Priced at a competitive Rs 289,000 against the Maruti Zen's 345,000, consumers flocked to the showrooms on 23 September 1998, when the Santro went on sale. Instead of a booking system, the Santro was made available against full payment. At the same time, Subbu's marketing team had already set up an extensive dealer network, with proper after-sales facilities. Whilst considering potential dealers, Hyundai looked at educational background, details of income, taxes paid and the kind of investment they could make. An important prerequisite for dealership was a fully functional workshop with properly trained people.

Moreover, Hyundai's marketing blitz was backed by deep resources and good manufacturing fundamentals. Hyundai Motor's investment of Rs 2,300 crore by 1998, was already the single largest by an automobile company in India till then. More importantly, the plant was not another assembly operation for knocked-down kits but an extensively integrated manufacturing facility. The Santros that were rolling out of the factory were cars that were being manufactured by the plant, unlike the assembly operations of Daewoo for the Matiz, or the way Maruti had begun the roll-out of most of their models till then — by assembling completely or semi knocked-down kits.

Hyundai had set up an aluminium foundry on site for cylinder heads and gearbox castings, with both engines and gearboxes made in-house from day one, in sharp contrast to Maruti Udyog, which kept importing gearboxes for years, till as late as 2006. Local content was over 70 per cent when production of the Santro started, a figure that many had not achieved till their fifth year of operation. By 2000, Hyundai Motor India had got to 85-90 per cent localisation. The very essence of Hyundai's strategy was to bring costs down right from the start of production by localising heavily, minimising losses, making the cars more price and cost competitive, and in the process, breaking even earlier than most other car-makers.

Of course, volume was necessary to meet the localisation targets, so Hyundai set up its new factory with an initial installed capacity of 100,000 cars per annum. At the same time, Hyundai brought along the Korean developmental philosophy right from the start: cars needed to be exported too. From the third year onwards, a target was set to export up to 40 per cent of production and 10,000 sets of transmission gears. For exports, the cars needed to be technically up-to-date, so Hyundai brought engine technology to match European requirements for emissions. With pollution becoming an issue in India, specifically in Delhi, the Supreme Court took matters in hand. Hyundai was ready with cars that met the emission requirements the Supreme Court believed should be introduced — standards set by the Euro II norms, or Bharat Stage II as it was named in India. Hyundai went to town to stress that its cars met Bharat Stage II, ahead of the rest of the Indian industry, thereby hinting these cars were technically more modern than those from the competition.

But it was the Shahrukh Khan campaign that worked really brilliantly. By August 1999, Hyundai had managed to shift 42,283 cars, and by the financial year-ending 31 March 1999, Hyundai Motor India had become India's second largest car-maker, after Maruti Udyog.

Hyundai set up a state-of-the-art factory in Chennai. Courtesy Hyundai Motor

By 27 April 2000, the 100,000th Hyundai rolled out of the Chennai factory, and by 12 June of the same year, the 100,000th Santro rolled out (Hyundai had launched the mid-sized Accent model too, in the meantime). A little over a month later, the first batch of 760 Santros and Accents were shipped out as the company's first serious export out of India (19 Santros had earlier been sent to Nepal in 1999, as the first tentative toe-testing).

A year later, on 11 July 2001, Hyundai unveiled a newer looking Santro, with a mild facelift. By then, the car-maker had made its 200,000th car in India. The 300,000th car rolled out a year later, on 26 June 2002, and the 400,000th on 8 May 2003.

It was on 22 May 2003 that an important event happened: Hyundai Motor India unveiled the Hyundai Santro Xing. It was an extensively reworked and facelifted Santro, with the front end completely redesigned as a markedly better looking car: Hyundai's ugly duckling had, after five years, become an almost-pretty cygnet.

Hyundai's Santro Xing was exported to several markets as the Amica.
Courtesy Hyundai Motors

More importantly, the redesign was not just for the Indian market, but was developed for all markets internationally. The Atos/Atoz had met with, at best, modest success in Europe, and sales in South Korea's domestic market was rather poor as Koreans preferred bigger saloons (the executive-sized Sonata had become South Korea's bestseller). So Hyundai decided that instead of two production bases, the Atos/Santro would be best served, in terms of scales of operations, by being made out of one factory — Hyundai's Chennai plant, to serve both its single biggest market, India, and for sales in Europe and the rest of the world.

The Santro Xing became Hyundai's global small car, and exports began in significant numbers. The Xing badged as the Atos Prime in Europe and most markets, Amica in the UK, Dodge Atos in Mexico, Inokom Atos in Malaysia and as the Kia Visto in Indonesia and South Korea. The first batch of 1,500 Santro Xings left for Europe on 12 August 2003. For India, the positioning changed. Initially, the Santro was positioned as an upgrade for Maruti 800 users, a complete family car for those who wanted more space and desired a bigger car. With the Santro Xing, the positioning changed, appealing to first-time car buyers.

Again, an innovative ad campaign played an important role in the transition from an upgrade to becoming an essential, the bread-and-butter car. A series of teaser ads on television had brand ambassador Shahrukh Khan surrendering the keys of a Hyundai Santro to Mr Kim, with the audience wondering whether Shahrukh's contract with Hyundai Motor had come to an end. Amid speculation that Hyundai was going in for a new brand ambassador, Mr Kim returned with a new Santro, handing over the key to a pleasantly surprised Shahrukh. Sure enough, the actor grabbed the keys and zoomed off in the Santro Xing, with Mr Kim left behind on the road...

With the launch of the Santro Xing, Hyundai took the car and brand to the smaller towns, along with opening over a thousand new service points across the country. Hyundai Motors backed that up by tying up with Indian Oil Corporation and Bharat Petroleum Corporation to position trained technicians at petrol pumps belonging to these two oil majors, extending service and repair facilities across the length and breadth of India.

By the end of 2003, Hyundai had made its half-millionth car in India and by end-October 2004, had exported its 100,000th vehicle — mostly Santros. The next 100,000 cars were exported in less than a year (by 18 October 2005). By 13 March 2006, Hyundai Motor India had rolled out its millionth car from its Chennai plant. By November 2006, less than a decade since its launch, Hyundai Motors India had made over a million Santros!

Interestingly, Hyundai got to its millionth Santro three years earlier than the Maruti 800, the only other Indian 'millionaire' car till then. Of course, the total car-buying market had grown exponentially since the launch of the 800, but then the Santro also had significantly greater and more redoubtable competition. Nor did it have the protective 'umbrella' of governmental largesse. Without doubt the product was intrinsically good, even if the styling was not. The Santro was very reliable, it was a practical car and it represented great value for money. Plus, Hyundai had its basics in place in terms of localisation and volumes (aided by exports). Servicing quality improved significantly with a dealership expansion programme in 2007-08. The number of dealers went up from 187 in 2006 to 320 by 2010. But there is no doubt the car's marketing and the way B V R Subbu and Hyundai played their media cards, contributed majorly to the Santro's success.

As late as 2007, when Shahrukh Khan was hosting the very popular television show *Kaun Banega Crorepati* (the Indian equivalent of *Who Wants to be A Millionaire*), during the 26 March episode, things took a dramatic turn when Shahrukh announced that one of the nine contestants who had made it to the hot seat would be getting a Hyundai Santro and the other eight consolation prizes. The consolation prize was a box, but within the box was a set of keys... to a Hyundai Santro, for each of the eight...The audience as well as the participants were shocked and overjoyed. The media, went to town the next day, and how! The stunt (initiated by a veteran journalist-turned-communication head), was another scoop for Hyundai.

But the Santro was soon to have an in-house rival. On 31 October 2007, Hyundai Motor unveiled the i10, the car that would replace the Atos Prime/Amica and most of the Santro Xing clones across the globe. But not in India. Hyundai decided to retain the Santro Xing in its range and position the i10 in a slot above, as a more modern, funkier-looking car from the same size segment. As the Hyundai i10 picked up steam in the Indian marketplace, the Santro Xing kept pace. Of course, in the export markets, mainly Europe, it was the i10 that was setting the sales charts afire, though the Santro Xing kept getting shipped to other markets, as Hyundai Motor India exported its half millionth car by end March 2008 and its millionth by February 2010.

The Hyundai i10, though planned as a replacement for the Santro, has not quite replaced Hyundai's first people's car. Courtesy Hyundai Motors

On 13 October 2011, Hyundai Motor India launched the Eon, the low-cost car designed to take on the mantle of Hyundai's price-leader, the car that would be Hyundai's 'people's car'. Priced lower than the Santro Xing, with a smaller three-cylinder engine, the Eon was decidedly more modern and funkier in shape, echoing Hyundai's more recent design language ('fluidic' design). The car looked very promising. But the driveability of the Eon, along with a somewhat polarising design, had buyers underwhelmed.

The Eon continues selling in increasingly decent numbers, but it has not quite replaced the Santro, as the 15-year-old veteran (at the time of writing in 2013), is still going strong, having sold 48,989 units in the calendar year 2012, and averaging sales of over 4,000 units every month till May 2013. The Santro has remained about 15th in rank among India's bestselling cars, despite being 'outflanked' by the i10 (positioned above) and the Eon (a notch below), which were, in 2012, India's sixth and eighth bestselling cars (with a combined total of 192,280 units sold). The i10 is another 'millionaire' wannabe and has chalked up sales of over 700,000 in India (and well over a million globally, with total sales at 1.46 million by mid-2013). Till mid-2013, Hyundai had sold over 1.27 million Santros in India and well over half a million had been exported globally. With well over 1.8 million Santros made, the car remains one of the most important cars not just in India but in the world. As production continues, it may just top the two million mark.

The Eon, Hyundai's people's car for the new millennium. Courtesy Hyundai Motors

Once the backbone of Hyundai's drive into India, the Santro today represents just a tad over 12 per cent of the South Korean giant's car sales in the country, which totalled 391,276 in 2012, keeping the car-maker at India's number two place behind Maruti. Yet there is no denying the important role the Santro has played, not just for Hyundai in India, but for the car-maker's success story and current place amongst the top five car-makers worldwide.

Hyundai's success story is remarkable when one remembers that by the late 1990s, Hyundai was short of air on American territory, struggling for one last gasp. Instead of retreating, as many others have done, the company made massive investments in new design and technology, other than addressing quality issues on a war footing. It also bought out Korean rival Kia, which, like Daewoo, was floundering soon after the Asian financial crisis of the late 1990s. And over the next decade or so, it integrated its design and sourcing activities with Kia's, though it retained the distinctive branding differences.

Hyundai's biggest rival in the 1990s, Daewoo, collapsed by 2000, and the car division was eventually taken over by erstwhile collaborator GM, which, in the irony of things, had been denied a majority stake in the 1970s by the South Korean Government. GM Korea is now South Korea's second biggest car-maker after Hyundai-Kia and most of the lower priced Chevrolets come from its facilities in Korea (the cars are still sold as Daewoos in Korea).

In India, GM India dropped the Opel brand eventually, to replace it with Chevrolet, and amusingly, most of the Chevrolets sold in India are models sourced from GM Korea.

In another strange twist of history, a bankrupt Daewoo's truck division was eventually taken over by Tata Motors in 2002 and renamed Tata Daewoo, with the trucks selling in Korea and several other markets as Daewoo or Tata-Daewoo.

Asia Motors became a part of Kia before the latter became a part of Hyundai. And as Asia Motors disappeared into history, a new commercial vehicle manufacturer was established in India, in 2002, called Asia MotorWorks Ltd (AMW), which had nothing to do with Asia Motors from South Korea...

SsangYong had a convoluted history too, having been taken over by Daewoo for a brief period, followed by Chinese ownership when Chinese car-maker SAIC took over SsangYong for a few years, before putting the latter into receivership in 2009. And guess who picked up SsangYong? None other than the other Indian automobile major, Mahindra & Mahindra, when in February 2011, it won over SsangYong for 522.5 billion won.

So South Korea's auto industry, other than the giant Hyundai-Kia conglomerate, is today made up of two (majority) Indian-owned automobile manufacturers, one American-owned car-maker, and a company called Renault Samsung Motors. Established in 1994, as Samsung Motors, by the white goods maker Samsung, the fledgling car-maker was soon in trouble thanks to the Asian financial crisis, when it was taken over by French giant Renault, which had, at around the same time, taken a controlling stake of Samsung's Japanese collaborator Nissan Motors.

In the year 2012, South Korea produced a little over 4.5 million cars and commercial vehicles, retaining its rank as the world's fifth biggest auto industry. But guess which country was close behind at sixth? India, with a little over 4.1 million vehicles produced. The other harbinger of a changing world order is that as many as 2.65 million passenger cars found buyers in India in 2012 as compared to a little over 1.4 million in South Korea; as the South Korean market contracts, or at best stabilises, India's is growing and is expected to continue doing so for the next few decades.

B S Seo, Managing Director and CEO of Hyundai Motor India, with the redoubtable brand ambassador Shahrukh Khan, at the roll out of the five millionth Hyundai – the brand new i10 Grand – at the company's Chennai plant. Courtesy Hyundai Motor

Of course, South Korea is still a major exporter of automobiles, whereas the Indian industry concentrates mainly on the domestic market. But only one of South Korea's six car-makers survives today as an independent Korean-owned entity. That one, Hyundai Motors, is not only one of the world's five biggest, but also one of the healthiest in an ailing industry.

Hyundai's success story is worth emulating today, more so than those of Toyota, Honda or Volkswagen. It has become the template for most new car-makers, with many of the Chinese upstarts acknowledging the 'Hyundai Way' is the way to go: get into the car business by making cars under licence, then gradually develop a range using mechanical elements from more advanced collaborators, and getting newer cars designed by Italian design houses on well-proven mechanicals.

Chapter VII
Playing the Alto

The unveiling of the Tata Indica, the Hyundai Santro and the ill-fated Daewoo Matiz in 1998, at New Delhi's Auto Expo, galvanized India's then dominant car-maker, Maruti Udyog, into some serious action — this, despite the fact that in the financial year 1997-98, Maruti's market share was already a staggering 82.7 per cent. The management at Maruti Udyog realised the good times were not going to last and they needed to do something, to react and take on the oncoming onslaught. But before that, Maruti Udyog needed to get its own house in order.

On 2 June1992, four months short of ten years of signing the original joint venture agreement between Suzuki Motor and Government of India, Suzuki increased its stake in Maruti Udyog from 40 per cent to 50 per cent. With that, Maruti Udyog ceased to be a national car company. This was fine in principle with most people, except that some industry experts, including Dr V Krishnamurthy, believed Maruti Udyog had been as good as gifted to Suzuki Motor... Suzuki paid just Rs 269 per share, a valuation that most, in retrospect, agreed was woefully below what it ought to have been. Moreover, there was the issue of principles — considering Maruti Udyog had been created in the first instance as India's national car project, it was very important to retain that aspect. Krishnamurthy's suggestion that, "Yes, Maruti Udyog needed to be privatised, but not by selling out to Suzuki, but by the government selling off 20 per cent of its 60 per cent stake to financial institutions, vendors, dealers and employees, so that it would be a board-managed company".

What Krishnamurthy had in mind was a car-maker not unlike Renault and Volkswagen in France and Germany respectively, both of which were successful Government-owned but board-managed companies. Both were being steadily privatised, without any single family or shareholder taking full control; forerunners of what was happening in China over the last two decades, where all car-makers were joint ventures with the local or Chinese Government, with not a single multinational having a controlling stake.

Of course, Osamu Suzuki, by then, had realised that Maruti Udyog was the ultimate cash cow. He was rather insistent on increasing Suzuki's stake to 50 per cent, which is what R C Bhargava, as Managing Director, pushed for too. This is what was 'gifted' to Suzuki. Virtually nobody protested; with the first waft of real liberalisation in the air, nobody dared question the decision, as it would have been seen as non-progressive and anti-liberalisation; out of place with the emerging winds of change.

The financial year 1999-2000 was good for Maruti Udyog: For the first time it made more than 400,000 cars. But overall car sales were down the following financial year — a 12 per cent drop — thanks to a recession. Maruti had unsold inventories and, for the first time in its history, went on to post a Rs 269.4 crore loss. That was followed by Maruti's first serious labour dispute.

All these problems galvanised Suzuki and Maruti into drastic cost-cutting measures, including offering a lucrative voluntary retirement scheme to workers and some managers. With 19 per cent of the employees leaving, Maruti Udyog became a trimmer, leaner set-up. Also, a rationalisation of vendors — addressing the problem of too many vendors chasing too few numbers — had Maruti trimming its supplier base from 400-plus to around 60-70. Maruti also decided to focus increasingly on first-tier vendors, buying aggregates and sub-assemblies from them, instead of sourcing each and every component separately from various sources. In the process, other than reducing costs, the car-maker benefitted from being able to maintain smaller inventories and dealing with fewer people.

At the same time, Maruti embarked on an extensive model renewal programme. The mainstay of Maruti's product line-up till then had always been the 800, introduced, as we know, as long back as August 1986. The Omni Van introduced in 1984, was a success story too, but the Gypsy (1985) and the Maruti 1000 (from 1990), had, at best, modest sales.

The launch of the Zen in 1993 was an important move for Maruti Udyog — it was to be the car that was designed to replace the 800 in the export market, and eventually supplement the 800 as Maruti's mainstay in the domestic market too. But it never did replace the 800, which remained Maruti's bestseller. It was only in 1999, with the launch of the Baleno and then the WagonR, that Maruti Udyog began to broaden its product offering from the earlier five-car model range to an unprecedented seven-car range that covered more than three-quarters of India's car market.

But the launch of the Alto on 27 September 2000, was the move that signalled the crossing over into the second phase of Maruti Udyog's rapid expansion programme. The Alto was Maruti's third try at making a 'people's car'. The first had been a huge success but it was ageing. The second, the Zen, was a failure and had been recently bettered and battered by both the Tata Indica and Hyundai Santro. Maruti needed an all-new people's car to take the fight back to the two new wannabes.

The Alto, Maruti's people's car once again. Courtesy Maruti Suzuki Ltd

Maruti Udyog's first move was the launch of the WagonR in India. A *kei jidosha* (literally, 'light automobile'), just like the 800 and the Zen, the WagonR made the most of the kei car restrictions: here was a car that stayed within the rules of the game in the sense that its length and width were those of a typical kei car, and so was the engine size. But by making it very tall and the seating higher, Suzuki managed to liberate a whole lot of space within. The WagonR (based on the same platform as the Maruti Zen) had big-car volume in a small car's dimensions — an idea the Japanese car buyers took to in a big way, making the WagonR a huge success in Japan.

The Hyundai Atos was one of many cars that were inspired by the thinking of the WagonR, and Hyundai's decision to bring in the Atos as the Santro, to India, must have triggered Suzuki and Maruti into believing the WagonR could take on the Santro. Which the WagonR did not really do, as the initial reception to the car's unusual design was, at best, lukewarm.

By 1999, Suzuki Motor was acknowledged as the international powerhouse for small cars, with a range of kei cars of all shapes and sizes, cars that were selling not just with the Suzuki badging but also as Mazdas, Subarus, Nissans, Opels and Vauxhalls, as the Japanese small car specialist supplemented ranges and supplied small runabouts to several international giants. Clearly, Maruti had a wide palette to choose from. And so it chose another car, the fifth-generation Alto from the Suzuki range, to take over the mantle of the 'people's car' from the Maruti 800.

Suzuki's fifth generation Alto, and its eighth generation mainstream kei car model, internally coded as HA12/22, was unveiled in October 1998, at the Tokyo Motor Show. Like the preceding Altos, the HA12/22 was another evolution, with the styling a little more rounded, where, without making the car too high, Suzuki had managed to carve out a tad more space within, compared to the 800 and the Zen. For Japan, as usual, a choice of 660cc engines, with or without turbocharging, was on offer.

In India, Suzuki decided to launch the HA12/22 as the Maruti Alto. Interestingly, it was the first time a Maruti was badged as the Alto — the moniker that had been used for the Zen in Europe, as well as for all the Suzuki small cars since the SS80, from 1980. For India, the decision was taken to launch the Alto with two engine variants: a base version with the three-cylinder 796cc engine as used in the Maruti 800, and a more luxurious 64bhp four-cylinder 1061cc engine version.

Though well received, the initial sales were slow, and the Maruti Alto wasn't exactly setting any sales charts afire. The 800 still reigned supreme, and there were takers for the Zen as well as the WagonR. In the first six months after the launch of the Alto in September 2000, till the financial year ending of 31 March 2001, Maruti managed to shift a little over 24,000 Altos. For the financial year that followed — April 2001-March 2002 — just 26,961 Altos sold in India; for the year after, sales came down to a most ordinary 25,771, averaging slightly over 2,000 cars a month.

True, the Alto had taken over from the Zen in export markets, selling as the Suzuki Alto across Europe and across the globe, but the going in India was very slow. The choice in India had widened considerably. Beyond Maruti there was Hyundai, Tata, Fiat and others. Though the numbers kept going up, Maruti was rapidly losing market share: by 2002, it had come down to less than a humbling 50 per cent. And profits too, kept falling.

Maruti Udyog needed to do something. So it did what it had been saying it would not do — slash prices. Beginning 2003, prices for several products were cut, the most important being the Alto, with Maruti offering a basic, no-frills version, sans air-conditioning, which took the car's price range 'down' to 800 territory. In turn, the 800's top-end version, the Dx, was discontinued. Yet the Maruti 800 still did very well, selling 20,000 units in March 2003. In fact, for the calendar year 2003, the Maruti 800 found 170,624 buyers, still outselling the Alto by a huge margin. Meanwhile, the Alto managed to climb up to a respectable 42,161 units that year, most of which sold after September, when prices had been chopped by Rs 23,000.

In the meantime, back in February 2001, the Government had decided Maruti would be listed on the stock exchanges and that it would gradually divest its stakeholding in the car-maker. Finally on 30 May 2002, following a rights issue, the Government reduced its stake, with Suzuki Motor taking full control with 54.2 per cent. In June 2003, Maruti Udyog went in for an IPO (initial public offer, in which the issue was oversubscribed 11.2 times) and by July, Maruti was listed on both the Bombay Stock Exchange (BSE) and the National Stock Exchange (NSE).

OUTING THE CHAMPION

The year 2004 was pivotal for Maruti. With further price cuts for the Alto, sales began to move north. In May 2004, for the first time, the Alto outsold the Maruti 800 — 10,373 versus 10,016 — and dethroned the close-to two-decade champion. In six more months, the Alto outsold the 800; yet, for the full year, the 800 was still India's bestseller at 132,706 sold, as against the Alto's 112,048.

For the financial year ending March 2005, Maruti Udyog crossed the half-a-million-car mark for the first time, making as many as 540,415 cars that year, thanks to the Alto and the 800, the new champ and the outgoing one. In fiscal 2004-2005, the Alto became India's bestselling car at 126,233, the 800 trailing closely at 116,262.

Without doubt, aggressive pricing played an important role. But there was no denying the Alto was the natural successor to the 800, given that it had all the plus points that had endeared the 800 to the Indian public, along with attractive additional features such as power steering, power brakes and a fifth gear. Also, with the Indian Government's move to encourage small car manufacturing, with concessional excise duty for cars under four metres and engine sizes below 1200cc for petrol-engine cars (and blow 1500cc for diesel-engine cars), the Alto benefitted significantly, as prices dropped further. And the numbers kept rising.

By the financial year April 2006-March 2007, the Maruti Alto continued to dominate the Indian car market, selling over 200,000 units for the first time. Sadly, the 800 dropped to a not so impressive sixth place, with sales at a little over 79,000 cars that fiscal. But Maruti managed to push the unloved WagonR to fourth, with the evergreen Omni at fifth. The second and third spots were commandeered by the Hyundai Santro and Tata Indica respectively. During the financial year 2007-2008, the Alto sold a record-breaking 227,193 units; most significantly, the Alto found its first millionth domestic buyer in November 2008 (after manufacturing its first millionth Alto in February 2008), a little over eight years since its launch. The first millionth 800 had taken 11 years.... The Alto's next millionth customer came in less than four years, as the second millionth Alto rolled out of Maruti's factory in May 2012. On the way, the Alto blitz ripped through the Indian marketplace, with sales topping an amazing 346,840 during the financial year 2010-2011, an average of close to 29,000 cars a month.

With sales rising and capacity at straining point, Suzuki decided it needed a second plant in India and a new factory was set up at Manesar, 20 kilometres from Maruti's straining-at-the-seams Gurgaon factory. The Manesar plant became Maruti Suzuki Automobiles India Ltd (MSAIL), created in 2004 as a subsidiary, with Suzuki taking a 30 per cent stake and Maruti investing 70 per cent. With complications in corporate governance, it was deemed best to merge MSAIL with Maruti Udyog Ltd in 2006. A year later, Maruti Udyog was renamed Maruti Suzuki India Ltd (MSIL), with both the Gurgaon and Manesar plants under this entity.

Maruti Suzuki's Manesar plant. Courtesy Maruti Suzuki

By 2010, the Alto had evolved too. The 796cc engine version remained the mainstay but the more expensive 1061cc engine was replaced in August 2010 by an Alto, with a newer power-pack. The Alto K10, as it was called, featured a new 998cc four-cylinder engine that was not only more powerful at 67bhp and 90Nm of torque, but was significantly more fuel efficient and cleaner in emission terms. The result was a peppier Alto that was not only quicker but was also more economical to run. To differentiate the Alto K10 from the more plebian version, the car had a redesigned front end, with a more sculpted pair of headlamps.

The K10B engine the Alto used was part of a brand new family of engines that Maruti Suzuki had started manufacturing since 2008, for use in its newer range of cars. An all-aluminium unit, the engine was lighter too, other than being more efficient, and could be made in several sizes. A bigger version of the engine — named K12M — in 1197cc displacement, was introduced for Maruti Suzuki's bigger cars; the smaller displacement version went into cars that needed to be exported to markets like Europe, where emission requirements were getting stricter. Though the Alto with the smaller engine remained the big seller, the spunkier Alto K10 found a new following, stimulating sales as well.

The Alto with a new heart and facelifted front became the Alto K10 – another success story. Courtesy Maruti Suzuki

In 2010, the Alto sold over 300,000 cars in India. That was when it was crowned the world's bestselling small car, a feat it repeated in 2011, when 311,367 Altos sold. The same year, Volkswagen's small car, the Gol, found 293,454 buyers in Brazil and Fiat's Uno sold as many as 273,537 units in the same country. But in 2012, the market slowed down for the Alto, with just 286,833 changing hands in India, even as the overall market grew by 14 percentage points to a record 2.65 million.

While all this was happening, Maruti launched the Swift in 2005. That marked a turning point in the car-maker's history. Till then, Maruti had been launching competent, reliable, workaday cars that made sensible buys; cars that were essentially everyday runabouts, nice but not necessarily very desirable. Cars like the 800 and the Alto. It also became obvious that what could be a people's favourite in another country need not be popular in the Indian market — the WagonR was a case in point. Japan's bestseller was criticised for its looks and it took a while to get the market to accept it.

The Swift was quite unlike any other Suzuki. It was stylish, funky, with attitude, and the ad campaign that ushered in the car's launch did wonders in making the car desirable — changing the image of Maruti with the targeted audience of a younger, smarter set. With the success of the Swift, the average age of the Maruti customer came down from 38 to 32. And the launches of several subsequent models — the SX4, the Dzire, the A-Star and the Ritz — helped bolster this newer, younger image.

The Swift, Suzuki's first really funky cult car. Courtesy Maruti Suzuki

But Maruti did not get it right all the time. Following the Swift, Maruti Suzuki launched the Zen Estilo, in November 2006; a 'tall boy' design that had very little to do with the earlier gen Zen. More WagonR-like in being high and spacious, yet arguably better looking, the initial excitement for the car faded away once the motoring press took the product apart for its shoddy on-road dynamics.

It was in January 2008, at the ninth edition of New Delhi's motor show, the Auto Expo, that Maruti Suzuki unveiled its first 'homegrown' concept car:

the A-Star. A very funky little five-door hatchback that previewed Maruti's small car replacement of the Alto for the export markets, the A-Star had everyone excited. Here was a car with striking good looks that promised a cheap but very stylish set of wheels for those who could not afford the premium hatchbacks but wanted something that reflected status.

The Maruti A-Star, a disappointment compared to the concept version previewed earlier.
Courtesy Maruti Suzuki

The production version of the A-Star was launched in December 2008, but the design, along with the tight packaging, fell far short of expectations. The car produced was not the design previewed, the original concept being significantly better looking. In a sense it had misled the public into believing the A-Star was going to be one good looker. Plus, the car was a little too pricey. The A-Star did manage to replace the Alto successfully for the European market (where, at the time of writing, it is still sold as the Suzuki Alto and rebadged, in slightly restyled form, as the Nissan Pixo), and till the end of 2013, was Maruti Suzuki's main export. But it has barely dented the sales of the Maruti Alto in the Indian market.

The replacement of the Maruti Alto, the car that could make Maruti the king again, finally arrived on 15 October 2012. Appropriately, it was named Maruti Alto 800.

The Maruti Alto 800, the next gen 'people's car'. Courtesy Maruti Suzuki

Developed entirely in India, with 200-odd Maruti and Suzuki engineers working on the project, the Alto 800 was not an all-new car. It was an extensively re-skinned and reworked Alto, but intelligently done to make a good car even better, yet retaining the essentials and keeping costs down. The car was developed not just for India but other emerging markets too like the African and Latin American.

The redesign was evolutionary, retaining the hard-points of the earlier car; the dimensions and location of the various pillars remained where they were in the Alto, with somewhat similar door opening, yet the designers managed to carve out a little more space within by raising the roof slightly and dishing in the seat-backs of the front two seats. By re-engineering the floorpan of the car, the torsional rigidity of the body increased, so the car twisted and flexed less than the Alto. Yet the weight of the Alto 800 came down by reworking several components.

Though the styling may not have been to everyone's liking, the Alto 800 was clearly more modern than the Alto, without being radically different, which was a 'comfort' factor for India's somewhat conservative middle-class and target segment Maruti knew only too well. The interiors, though, were a significant improvement on the Alto. With intelligent use of low-cost two-toning, the interior managed to look more cheerful than most of the

Alto 800's rivals.The most interesting feature though, was the last bit of the moniker — the simple fact that the India's next gen 'people's car' was, once again, an 800 — using the same basic powerpack that started it all three decades ago. True, nitpickers will point out that the latest version of the 796cc engine was not quite the same as the contraption that ignited India's automotive revolution back in 1983, but there is no denying it was a nicely evolved version of Suzuki's F8B engine, dating back to July 1981!

The bore and stroke of the engine, the displacement (or swept volume), were identical. The top end of the engine changed — instead of two valves for each of the three cylinders, there were now four valves, two for the intake and two for the exhaust, thereby improving the 'breathing' of the engine, and thus the efficiency of the motor. Power and fuel efficiency and, thereby, 'cleaner' exhaust emission, was further improved by the use of a sophisticated multi-point fuel injection system instead of the ancient carburettor of the old engine. Otherwise, the basic 'architecture' remained the same, even if the nomenclaturing has changed subtly to F8*D* (as it was two generations more modern than the original).

An interesting reminder is that a little over two decades ago, Suzuki convinced the Maruti management that there was not much life left in the F8B and that it needed to invest in the Zen if export ambitions were to be realised. The F8B, incidentally, is soldiering on too (albeit in fuel injected form), as motive power for the evergreen Omni, which has also sold in excess of a million… Of course, it remains to be seen whether the new Alto 800 will be a 'millionaire' car, but if the first six months of 2013 was anything to go by, the Alto and the Alto 800 (with the latter outselling the former two-to-one), have reported combined sales of more than 137,000 units, indicating the Alto 800 could be the subject of a future chapter in a future edition of this book.

Whilst these model changes were taking place, there was also a change of guard at the top, with Shinzo Nakanishi becoming Maruti Suzuki's first Japanese Managing Director (with R C Bhargava back as part-time Chairman), in December 2007, when Jagdish Khattar retired. Year 2008, soon after Nakanishi took over, was a cataclysmic one for the auto industry worldwide, but the Indian market held steady, albeit growing marginally. Maruti Suzuki's sales for fiscal 2009 were barely 1.45 per cent better than fiscal 2008. However, the subsequent years have been much better.

RC Bhargava at the helm of affairs once again. Courtesy Maruti Suzuki

With Nakanishi taking over, Maruti Suzuki saw an evolution in its management structure, with a distinctive shift in style and attitude. Though, in terms of productivity and structure, Maruti Suzuki is moving towards the 'Suzuki' way and a more 'Japanese style', with better productivity, greater teamwork and decision-making; with responsibilities being shared equally between the senior Indian managers and their Japanese counterparts, Nakanishi insists the car-maker will retain its 'Maruti style of management', which he thinks is a nice mesh of the best of Western, Indian and Japanese work practices and ethos. Also unusual is the top management structure of paired leadership, wherein there is a Japanese executive for each Indian in top positions. There are five such pairs.

Other than the obvious adding of capacity, a chunk of Maruti Suzuki's future spending will be on developing its research and development (R&D) facilities, which has been minimal till fairly recently. Maruti does not need to do this as there is its joint venture partner Suzuki, always there, ready to supply it with products or support in terms of product modifications and changes. But more significantly, Suzuki did not want to 'empower' Maruti as long as it did not have full control; the Japanese have always been reluctant to encourage indigenous abilities….

Though I V Rao, the former Managing Executive Officer (Engineering), points out that, "…product development activity started when we first modified the interiors of the Van by designing the new rear seat", way back in the 1980s, the real learning came from the Zen facelift, followed by the WagonR facelift programmes between 2003 and 2006. Also, the realisation that, "…a car doing well in Japan would not necessarily do well in India", prompted Maruti and Suzuki to develop a semi-independent R&D centre that would cater specifically to more distinctive Indian tastes.

A good example of that was the R3 concept that Maruti Suzuki showcased at the 2010 edition of the Auto Expo in Delhi — a multipurpose van (MPV) that was being specifically configured for India (and Indonesia), and allowed for 5+2 seats in a compact package without looking too much like a boxy van. The Maruti Suzuki Ertiga MPV, launched in 2012, was the production version of the R3. The next design milestone was the Alto 800, an example of extensive body redesign. It is believed Maruti Suzuki will have full capabilities for complete model development by 2016. This is prompted by the thinking that India is closer to Europe in design preferences but ultimately has its own tastes. It also stems from the thinking that, in future, styling will be as important as performance. Specifications will remain more or less the same between brands, with design being the main differentiator.

The Maruti Suzuki Ertiga multi-purpose van, launched in 2012, and the Alto 800, are harbingers of things to come in the sense that over the next few years, we can expect to see Maruti Suzuki cars that will not only cater to the specificities of the Indian market but those of different export markets, and not just in Europe. Increasingly, more cars from Maruti Suzuki will be sold in Africa, Latin America and Asia. Maruti Udyog Ltd, the nation's pride two decades ago, the maker of the first Indian 'people's car', has, over the years, become the single most important outpost of the Suzuki empire, and will, in the years to come, remain the main driver of the Indian automobile industry.

In more ways than one, the story of Maruti has been the story of the Indian automobile industry of the last three decades. It took Maruti Udyog Limited 11 years to produce a million cars. It took another four years for it to make its next million cars (by 1998). And then it had to wait another dozen-odd years for Maruti Suzuki India Limited to make a million cars in a year. That was during the financial year 2009-2010, when, in a first, an Indian car-maker manufactured a million cars in a year. Since then, it has been doing well over a million every year with 1.17 million vehicles made during the financial year 2012-2013.

Shipping of the millionth Maruti to international markets in April 2012.
Courtesy Maruti Suzuki

Of the 2.65 million cars sold in 2012 in India, Maruti Suzuki accounted for 1.06 million, straddling 40 per cent of India's car market, despite ever increasing competition. Since inception, Maruti Suzuki has sold over 10 million cars in India. Additionally, over half a million (700,000) cars made by Maruti Suzuki have been exported worldwide. To paraphrase the popular adage of General Motors: what's good for Maruti Suzuki could very well be good for India (even if it is not an 'Indian' company anymore). The days of Indians being at the helm seem to be over too. When Nakanishi retired in March 2013, his place was taken by another Japanese, Kenichi Ayukawa.

If Maruti had remained an independent car-maker, it could easily have ranked amongst the top 20 international majors, joining the millionaire car club (car-makers making over a million vehicles every year); rubbing shoulders with the likes of BMW and Volkswagen, and breathing down the neck of Mitsubishi Motors, which got into the car-making business 66 years before Maruti did.

Most noteworthy is the fact that since it became a subsidiary of Suzuki Motor Corporation in 2008, Maruti has helped Suzuki get into the list of top 10 major international car manufacturers, whereas, just a decade earlier, Suzuki was ranked 13th worldwide. At that point of time, sales of cars in India and exports from Maruti contributed to over a quarter of SMC's sales worldwide. Since then, Maruti's sales have zoomed, as traditional car markets across the globe — where Suzuki has always had a significant presence — imploded, taking Maruti Suzuki's share in Suzuki Motor's global sales figures to over 40 per cent, and propelling Suzuki to the ninth spot among car majors in the world.

Perhaps we need to rephrase the GM-US statement once again: what's good for Maruti Suzuki is indeed very, very good for Suzuki Motor Corporation.

Chapter VIII
Day of the Nano

The night before 10 January 2008, Ratan Tata, then Chairman of Tata Motors, and Ravi Kant, its Managing Director (and later Vice-Chairman), got little sleep. Tata, the perfectionist he is known to be, his senior management, and the young team that had developed the latest Tata Motors prototype, went through the minutest details regarding the unveiling of what was being touted as India's most important car ever — the '[Rs] One Lakh Car' — that by mid-day of 10 January, everyone would know as the 'Tata Nano'.

Yet Ratan Tata looked as fresh and healthy, belying his 70 years, when he and Ravi Kant walked in for a breakfast meeting with a select group of newspaper and magazine editors at eight o'clock sharp at the Longchamp banquet hall of the Taj Mahal Hotel in Lutyen's Delhi. It was time for a little debriefing, a summation of all that had been said during the preceding four years: yes, it was going to be a proper car; yes, the engine would be at the rear; yes, it could seat four, even five at a pinch; yes, it had four doors; yes, the name was under wraps for now; and no, no pictures were available before the unveiling later in the day.

Ratan Tata seemed in a somewhat combative mood, taking on the various criticisms that had been levelled just days before the unveiling. Issues like emissions and congestion were discounted, with Tata explaining the car would be cleaner than most cars in India, in fact, cleaner than many scooters and motorcycles, and even if Tata Motors managed to produce half a million cars every year, it would not account for even 2.5 per cent of the vehicle population of the country, let alone the globe and the One Lakh Car would not be the cause of worsening traffic congestion in cities like Delhi and Mumbai. Thus it was on a cheerful note that Ratan Tata, Ravi Kant and the other invitees headed for Pragati Maidan, Delhi's most important exhibition ground, where in Hall 11, at 11.30am., the much-awaited all-Indian peoples' car was to be unveiled.

The night before, a container vehicle carrying three cars had reversed into the side entrance of Hall 11, where these highly secret exhibits were unloaded in cloak-and-dagger fashion under cover of darkness. Tata Motors had as many as 17 other vehicles on display at its stand, and at least three other new vehicles were being unveiled for the first time (including the all-new replacement of Tata Motors' bestseller till then, the Indica Vista), yet the kind of security and secrecy surrounding these three little cars, was on an altogether different level. Either way, the mission was successful: the cars had been 'smuggled' in the evening before without any paparazzi managing to get even the haziest shot of the much awaited l'il car. On that cool wintry Thursday morning, nothing else mattered — none of the other cars, other launches, the leggy models…the entire focus was on getting into Hall 11 and if one succeeded, the unending wait for the official launch of the big little news of the day.

All the waiting spectators, journalists, industry-wallahs, two-bit politicians, minor celebrities and hangers-on, all of whom had managed to squeeze into the overcrowded Tata stand could do, was speculate on the name of the car. The latest buzz was that the car was to be called Jeh or Atom. Jeh was supposedly derived from the first three letters of Jehangir Ratanji Dadabhoy (JRD) Tata, Ratan's uncle and Chairman of the Tata group before Ratan Tata. In Persian, *jeh* meant 'conqueror of the world'. Surely, Tata's little car was going to conquer the world!

Though the press conference was supposed to start at 11.30am, the Tata stand was chock-a-block an hour before. By 10.45am, there were no vacant seats and TV crews had already taken positions, as photographers jostled for vantage points. Each time there was a buzz that Ratan Tata was about to arrive, the photographers went ballistic. The crowd was building up, excitement running high and the edgy mood palpable. Getting to Pragati Maidan was a job in itself as the traffic was unusually heavy while getting inside Hall 11 was getting more and more difficult as security men controlled the entrance, filtering invitees, journalists and industry folk. More than a thousand people packed the hall.

Day of the Nano

A few minutes before 11.30am, the lights were lowered, the floor was bathed in a blue-green light, and as the hushed mass of people waited for a Bollywood troupe to jump out and boogie to foot-stomping music, a hologram of a family on a scooter appeared instead — a man driving, a woman behind, with a child in her arms, and another child perched precariously on the footrest. The hologram faded out to be replaced by an image of Ratan Tata himself. As more than a thousand pairs of eyes watched in awed silence, Tata's voice boomed, "Ladies and Gentleman, thank you for being with us on this memorable occasion. There are no celebrities at this function nor any dance routines. The centre of attraction for this morning's event is the new Tata car which we are unveiling. We're going to take you on a small journey. A journey that symbolises the human spirit of change, the will to question the unquestionable, the drive to stretch the envelope. Ladies and gentleman, I invite you to join me in this journey of innovation and evolution. The quest to lead and the quest to conquer."

Tata went on to explain the background and genesis of the project: "Today's story started some years ago when I observed families riding on two-wheelers, the father driving a scooter, his young kid standing in front of him, his wife sitting behind him holding a baby and I asked myself whether one could conceive of a safe, affordable, all-weather form of transport for such a family. A vehicle that could be affordable and low cost enough to be within everyone's reach, a people's car, built to meet all safety standards, designed to meet or exceed emission norms and be low in pollution and high in fuel efficiency. This then was the dream we set ourselves to achieve. Many said this dream could not be achieved. Some scoffed at what we would produce, perhaps a vehicle comprising two scooters attached together or perhaps an unsafe rudimentary vehicle, a poor excuse for a car. Let me assure you and also assure our critics that the car we have designed and we will be presenting to you today will indeed meet all the current safety requirements of a modern day car. Anyway I have said enough ladies and gentlemen, now I give you the new car from Tata Motors, the people's car that everyone has been waiting for."

In a halo of lights, with the theme music from Stanley Kubrick's 1968 classic, '2001: A Space Odyssey', blaring in the background, as bits and pieces of the much-anticipated car flashed on the screen behind, a iridescent silver-white egg-like ball of metal emerged from the shadows, rolled onto centre-stage, and Ratan Tata himself stepped out. Within seconds two other cars, one red with black bumpers and the other a bright yellow, were driven onto the stage and parked, flanking the silver one.

Ratan Tata and Girish Wagh at the launch of the Nano. Courtesy Tata Motors

Thunderous applause…and then mayhem ensued as photographers, journalists, the audience, everyone scrambled for a closer look. In the midst of the excitement, Tata quipped, "Some people said we should call it the 'Buddha car', while others said that we should call it 'Mamata' or 'Despite Mamata'," drawing laughter. "We decided to call it 'Nano'… since it is high-tech and small. The car will be priced at Rs 1 lakh. A promise is a promise!"

The Tata Motors team at the launch of the Nano. Courtesy Tata Motors

Day of the Nano

Till late that evening, Hall 11 at Pragati Maidan remained jam-packed. Outside, the eager crowds reminded you of a cricket stadium before a one-day match. Hundreds of security men formed uncompromising barricades with thick ropes.

It wasn't just on January 10 that the crowds were crazy — on subsequent days of the Auto Expo, the hall that everyone was heading for was the one that had the 'lakhtakia' car. Despite the fact that there were more than two thousand exhibitors from over 25 countries displaying everything from automobiles to accessories (aided by leggy models) at the Auto Expo, thousands upon thousands came essentially to get a glimpse of the Tata Nano. By January 16, the last day of Delhi's 9th motor show, some 1.8 million people had thronged the Expo, comfortably beating the Paris motor show's draw of a million-and-a-half and perhaps eclipsing that of the Bangkok motor show too. The daily, *The Hindustan Times* (in its January 14 edition), reported: "They came in their thousands, from Delhi, Haryana and UP — riding cars, buses, the Metro, and even tractors and tongas. They filled Bhairon Marg and Mathura Road, and set off traffic snarls that stopped Delhi at several places."

For what everybody had gathered to see at Hall No 11 in Pragati Maidan was not just another small car, but to see hope emerge on wheels. For this 'lakhtakia' car, in Hindi meaning 'the one-lakh rupee' car — as the man on the street had already named it — had enabled millions to dream of a life beyond the motorbike, of a life that would be safer and more comfortable for themselves and his (or her) near and dear ones.

The media and the Nano

The morning after the launch, newspaper headlines about the 'lakhtakia' car screamed for attention — from a nicely said 'Small Wonder' to a more dramatic '1 lakh car drives 1 billion dreams'. Since the 1983 launch of the Maruti 800, no other car in the history of India had garnered as much newsprint space or as much airtime in India as the Tata Nano had. Every TV news channel in India and some of the most important news channel across the world — whether it was *CNN*, *BBC* or *TF1* in France — all had the Tata Nano as headline news on the evening of the 10th. The cheapest car ever in the history of the automobile had captured the imagination of the people, and not just in India, but across the globe.

While India's press went hyperbolic, the press across the globe was not necessarily that enthusiastic, though most did admit that the Nano was a very significant car. "Many were surprised by the Nano's natty styling, confounding pre-launch predictions that a car at that price would be little more than 'a super-charged autorickshaw' or 'two motorcycles joined at the hip'," said Peter Foster and Pallavi Malhotra, on London's *Daily Telegraph* website. Positive reactions were also expressed by other media, among them *Automotive News Europe*, *The Times* of London, *Forbes*, *The Economist* and *The Washington Post.*

The reaction from the industry in India, and worldwide was no less telling. While some Tata Motors rivals like Anand Mahindra of Mahindra & Mahindra, Venu Srinivasan of TVS Motors and Sylvain Bilaine of Renault India were all praise, others were somewhat sceptical.

Anand Mahindra, managing director of Mahindra & Mahindra (speaking before the unveiling) was quoted as saying: "I think it's a moment of history and I'm delighted an Indian company is leading the way." His colleague, Dr Pawan Goenka, the president of the automotive sector of Mahindra & Mahindra, said: "As an Indian, I feel proud that an Indian company has launched a product that the world is looking at. As a competitor, we think it is something that we will have to fight with as our three-wheeler's price overlaps with Tata's car."

Venu Srinivasan, the chairman and managing director of TVS Motors, said: "It is fantastic, outstanding engineering. It helps redefine the sector in the country. It has established new grounds." Similar sentiments were echoed by Sylvain Bilaine, the chairman of Renault India then, Arun Firodia, the chairman of Kinetic Engineering, Rajiv Bajaj, the managing director of Bajaj Auto and Jagdish Khattar, the former managing director of Maruti Suzuki. Many senior managers from the international car majors also had nice things to say. Ford's executive vice-president John Parker praised the car, saying: "It is a groundbreaking product. The Nano will cause people to think differently about the car. I have a lot of respect for Tata."

Not surprisingly, some of the competition had reservations, but other than questioning the car's features and abilities, most didn't criticise too much, except for Wang Chuan-Fu, the chairman of the Chinese car-maker BYD.

When questioned about the Nano at the Detroit motor show that followed a couple of days later, Wang said: "That car doesn't have air-conditioning, power steering, air bags and other features. Do you dare to buy that kind of car?"

The main criticism came from environmentalists and social activists. R K Pachauri, the chairman of the Inter-governmental Panel on Climate Change (IPCC), which organisation, along with Al Gore, was a recipient of the Nobel Peace Prize, said the Tata small car was giving him "nightmares", implying the environmental impact of emissions from more cars on the streets. Pachauri was quoted by the Mumbai-based newspaper *DNA* as saying: "This is not the transport option for a country of a billion people, many of whom cannot afford to buy even a bus ticket."

Similar concerns were apparently expressed by Sunita Narain, the director of the Centre for Science and Environment (CSE). But Narain was later quoted in *Business Standard* as clarifying that, "the small car is worth celebrating as it will help meet the aspirations of a large number of people. But it is not the same as meeting the needs of a large number of people," explaining that just 20 per cent of the people of Delhi, for instance, were users of two-wheelers and cars, whereas the rest had to depend upon an inefficient public transport system, and to her that was greater priority.

But then Dinesh Mohan, from the Indian Institute of Technology, Delhi, had a riposte to that: "Those who oppose it are like dogs in the manger who don't want everyone to have a car they themselves possess. If at all anyone wants clean air, let them give up their big cars and SUVs and their second and third cars."

In fact, there were quite a few sections of the Anglo-Saxon press and people who expressed great reservations. Mike Rutherford, the rather outspoken British columnist who writes for *The Times*, the *Daily Telegraph*, *The Independent* and the leading British weekly *Auto Express* said in a column: "For heaven's sake, Tata's main claims to fame on the automotive stage are that it makes unsophisticated buses and lorries, atrocious 4x4s and the world's cheapest car, which is not much more than an upholstered skateboard."

Across the pond, columnists from the *New York Times*, *Newsweek*, and several other media outlets blasted the Nano. *NYT* columnist Thomas Friedman —

a fan of India otherwise — wrote that the Nano was a "cheap copy of our worst habits". *Newsweek*, in an article headlined 'A Billion New Tailpipes' was far more critical: "It turned out to be a four-seater, a bit more than three meters long, with a 624cc engine and made of plastic and glue instead of welded steel."

The American dislike for the Nano spilled over to the internet too. Of course, the most interesting viewpoints came from bloggers. On 10 January 2008, the day the Nano was unveiled, the Tata Motors website had 7.9 million hits, while the Tata Nano website had four million in 30 hours, making these sites among the busiest in the world!

THE NET REACTION

I live in the land of ridiculously large vehicles, so large that the average vehicle is 1L per person for engine displacement, or bigger... we should be ashamed here in North America. Thank you for your inspiration, Mr Tata and keep up the good work. Smart, Nano, Festiva, Geo, Suzuki... I hope these are some of the names that will bring forth the next generation of transportation – Jean Marc

I want one. NJ stinks. I could throw that money on a credit card and show people a good reliable car and be the 1st in my town WITH ONNNNNNE! – Jonathan

...and I paid $18,000 for a motorcycle! – R Welsh

This car would be great for Americans, till you get crushed by a reckless SUV or cargo van – VCHAGZ

To Americans above who say it will produce a bunch of pollution, I can guarantee it takes 1,000 of these tiny cars to produce the pollution of one gas-drinking, 8,000 pound Detroit SUV. Quit whining. Congratulations to Tata Motors – Stan Buie

The least polluting (CO2) vehicle in the world was just launched (at 50mpg). ... with fuel costs going uphill everywhere this is a welcome addition to the marketplace. I have a gut feeling this thing (or one of its descendants/imitators) will dominate urban driving everywhere. Especially in India. That is one bodacious Tata!! – Carl

Had Ford or GM followed this model 25 years ago there would be no:

- *War in Iraq*
- *SUVs*
- *Soccer Moms in the darned SUVs*
- *President Bush II*
- *Impending war in Iran*
- *Reconstruction of the Communist Block*
- *Smog warnings.*
- *100 dollar barrels of oil*

But there would still be a lack of public transportation – Seth A

Without a doubt, the wave of the future, especially if the next version is electric or a hybrid. Brilliant idea, particularly for the billion-plus Chinese right next door – Karekin

This is the greatest achievement in the history of the automobile. How much to FedEx it to the US? – Albert Kunze

Amazing! How is it possible that this car costs less than the new Mac Pro computer?! – Dave

Go India Go!! It's truly a remarkable achievement. Hats off to Indian Engineers… P.S : I am a senior level executive working for Toyota Motors – William McDown

What the Citroen 2CV was to millions of French and Africans in the 1950s and 1960s, the Tata Nano will be to many parts of the world in the 2010s! We NEED this car in France – Raphael

Wow. I'm amazed how many people can be impressed by a backward piece of junk. As the United States begins to actually move away from oil dependence and dirty technologies, the Third World eagerly awaits its chance to be the new oil junkies. H L Mencken would still be right if he were alive today: "No one ever went broke underestimating the intelligence of the Indian public." – John

People's car? Hey, wait a minute! In German, that comes out 'Volkswagen'. This sure looks like a reinvented Beetle – 70 years after the original – Unimpressed

Day of the Nano

Without doubt, Ratan Tata and his team had orchestrated an extraordinary unveiling — an event that had not only captured the imagination of the teeming masses of India, but that of people across the globe. So could they, orchestrate yet another epoch-making event less than two months later, when the Nano was to be shown for the first time to the international media and audience at the Geneva Motor Show? By the start of the first press day, on 2 March 2008, there was no doubt the little Nano was going to be the big star of the show. Everybody had the same question: 'What time will the Tata Nano be unveiled?'

Ratan Tata and the Nano at the Geneva Motor Show. Courtesy Tata Motors

When, exactly 13:30 hours, the Tata press conference commenced, press veterans from across the globe thronged the stand, eager to take in every word Ratan Tata spoke and every sight of the two Nanos, one white and the other orange. They had caught the fancy of the even the Doubting Thomases in the ranks of the world motoring press.

Day of the Nano

The conference finished, Ratan Tata left, yet the crowds remained. The GM and Renault-Nissan supremos, Rick Wagoner and Carlos Ghosn, dropped by, checked out the Nano, and had interesting observations to make. Ghosn was all praise: "I should thank you, this is really fantastic. This is what we should try and achieve." Rick Wagoner opined the Nano was "a benchmark kind of model," adding, "...you Indians always make us think." Yet, even when they moved on, the crowds remained.

Tata and the Nano were the talking points of the day. The next day, many newspapers across Europe had the Nano on their front pages. The Nano had already become a star in the old continent. Clearly, Tata Motors had finally arrived on the global stage — the Nano had more than pushed the boundaries in a direction no other car-maker had ever imagined. And it was obvious that even among the most discerning and demanding Europeans, there were many who believed the Nano was the right "product at the right time," as Silvia Baruffaldi, Managing Editor of the Italian design magazine *Auto & Design*, said.

"This cute little car is interpreting the current trend of downsizing in a very smart way, and not just for its reduced dimensions, weight and fuel consumption, but also in its global project approach," explained Baruffaldi, "Tata's designers and engineers went back to the fundamentals, to avoid all the add-ons of modern cars in order to concentrate only on the essential parts, reducing the number of components and introducing modular design for many solutions. Style was not at all neglected in this less-is-more approach. A good balance of rounded lines makes the Nano look nice and pleasant, far away from the uninspiring boxy shapes of Japanese micro cars. There's even a hint of sportiness in the side air intakes for the rear-mounted engine cooling.

For the Indian car market, this car could act as Fiat's 600 and 500 did in Italy in the 1950s and the 1960s: friendly looking, compact and affordable vehicles which were able to give access to individual mobility to a vast number of users. The passenger compartment of the Nano in its entry level version is actually a modern interpretation of those cars' cabin, while the symmetrical layout of the dashboard, easily adaptable for both right- or left-hand drive is clearly intended for international sales."

Clearly, by mid-2008, the world wanted the Nano. The hoi polloi wanted one. The chic set wanted one too. From India to Indonesia, France to Zanzibar, everyone seemed to want the Nano. Yet things did not quite go according to plan or expectations. What went wrong? Was there a warning somewhere, sometime in 2008, which the people at Tata Motors did not listen to?

On paper, the car looked good. Or was that indeed so? What did the experts have to say at that point of time? What did those strange breed of men and women who review cars for a living, who are painfully critical, think of the car? Sure, the car had made headlines in newspapers, as much as it had caught the attention and fancy of the motoring press. Even if it did not make it to the cover of any of the significant international car magazines on its unveiling in Delhi, it was reported in every major car magazine across the globe. Of course, it was on the cover of every Indian car magazines in February 2008, and every magazine devoted several pages to the Nano. But the articles then were based on the specifications of the car and what the car looked like. No one knew what it was like to drive one. For that, the motoring hacks had to wait.

But in the meantime, a few articles really stood out: an article headlined 'So who is this Mr Tata?' in the February 2008 issue of France's leading car mag *l'Automobile*, which had the Nano as the centerpiece news point and made mention of Tata's plans to acquire Jaguar and Land Rover. Another noteworthy piece was the design analysis of the Nano in the June 2008 issue of the American monthly *Automobile*. The design of the Tata Nano was reviewed by design expert, Robert Cumberford, who has designed several cars himself over the last four decades and had many interesting observations to make: "Here it is at last, the $2500 car that Ratan Tata proposed several years ago, perhaps the most significant new car since the Ford Model T was introduced 100 years ago." France's leading automotive website, *leblogauto.com*, voted the Tata Nano the Car of the Year for 2008! Ben Oliver of British car magazine *CAR* said: "...the Nano is a staggeringly important, clever, exciting new thing."

Veteran auto journalist, Richard Feast, known for his outspokenness, had this to say: "I haven't been to a motor show, press conference or car launch in two years now. I don't read the car mags. All of which means I've never even seen a Nano. I'd read about it in the newspapers, so all my information is secondhand. Having said that, I think the concept is BRILLIANT. I hope it succeeds, if only to put two fingers up those smug western companies that said it couldn't be done, there was no market for it, etc, etc…." Accolades also came from Sam Hardy, the Deputy Motoring Editor of the UK's bestselling weekly car magazine, *Auto Express*, Frédéric Guihal, Associate Editor of the French car magazine, *Auto Moto*, Alexandra Legendre, associate editor of *l'Automobile,* and Stéphane Meunier, Editorial Director of Motorpress France, who said: "For me, the Nano is less of an alternative amongst automobiles: it is more of an ambassador for India, an ambassador that announces the new India of IT, steel, automobile, of the fact that Jaguar and Land Rover are part of the same group. The Nano is the Indian UFO that has caught the fancy of the media."

The thought was reflected by veteran Indian journalist and writer, Rahul Singh: "I have a list of defining moments when India, instead of being considered a perpetual basket case, started being looked at positively by the developed world. One, our nuclear tests. Two, at the risk of sounding flippant, our successes on the stage of beauty, that is, Sushmita Sen and Aishwarya Rai. Three, our stunning achievements in the Information Technology area. But if you were to ask me if there was one single event that really defined India's entry into the consciousness of the people who matter internationally, it has to be the launch of Tata's Nano. Nothing else so captured the world's imagination."

In some ways it was similar to a story another prominent journalist-columnist-writer, Anil Dharker, related in the summer of 2008: "The French Riviera is full of `old villages'. These are perfectly preserved remnants of older settlements adjoining the fashionable towns of Cannes, Nice and Antibes. For reasons unknown, most of them are uphill so there's a price to pay for doing something less touristy (Although truth be told, it's only tourists who go up to see these places). One morning we were trudging up a steep hill that was to take us to Eze. We heard the puttering of a lawn-mower type engine and turned around to see a motorized trolley laden with large bottles of Coke and other soft drinks. Guiding the trolley was a grizzled old Frenchman, who saw us, stopped and said, 'You are from India, no? You know Tata company? I am going to buy a Nano. The first car I will ever have.' His smile was as bright as the sun shining overhead. It struck me then that the Nano wasn't just an Indian phenomenon. It was an international one, a revolutionary idea that could have a world-wide impact in the years to come."

In retrospect, Dharker, Singh, Meunier, Legendre, all seem to have spoken too soon, as had many others. What went wrong despite the unprecedented acclaim and praise? With praise comes pride. And pride, more often than not, is followed by a mighty fall. With the unprecedented media coverage, the people involved lost their perspective. A certain hubris had begun to set in amongst the engineers and managers of Tata Motors as one triumph seemed to beget another.

Chapter IX
Drawing Board to Driveway

Though work on the tata nano project began in early 2003, the seed of the idea had been sown by Tata Motors way back in 1998, when the Indica was unveiled. At the 1998 Auto Expo, during the launch of Tata's first car, a tiny, semi-finished concept was also on display at Delhi's Pragati Maidan.

Tata's Xing concept from 1998. Courtesy BS Motoring

Passed over in the excitement and promise of the Indica, the yellow-and-black painted mock-up (a full-size scale model of the proposed shape), called the Zing, was a harbinger of sorts. That small cuboid shape — referred to as a one-box design in automotive parlance — predicted the shape of things to come. The design was that of a very compact four-door 'mono-volume' car, though the location of the engine was not clear at that stage.

Even if the Indica was to be the mainstay of the Tata Motors passenger car model for the following decade, somewhere, sometime, Ratan Tata and others at Tata Motors, were plotting their next big move: a car (or some form of personal transportation), that bridged the gap between two-wheelers and the cheapest car in India.

As production of the Indica and its variants settled in, work began on a cheaper automobile in late 2002. The then Tata Motors Executive Director, Dr V Sumantran, who was also the company's technical head at that point of time, was asked by Ratan Tata to explore the possibility of developing a vehicle that would be between "...a two-wheeler and the lowest priced car"; one that would "...enable the family to migrate from a scooter or motorcycle to a four-to-five-seat car".

"Mr Tata was also influenced by C K Prahalad's book, *The Fortune at the Bottom of the Pyramid: Eradicating Poverty through Profits*," remembers Sumantran. Prahalad's proposition, that businesses should stop thinking of the poor as victims and instead start seeing them as value-demanding consumers, had already driven several Indian and India-based multinationals such as Godrej and Hindustan Lever (now Hindustan Unilever), to launch ultra-small sachets of everything from shampoo to *gutka* (chewing tobacco), sparking off a retail revolution. Tata Motors saw a similar opportunity in the car market.

The first step was to form a core team that could work on studying and exploring possibilities. This Concept Evaluation Team (CET), made up of five members representing the various important disciplines of body engineering, packaging, safety regulations, styling and engine — namely, R G Rajhans, Jai Bolar, Anil Kumar, Nikhil Jadhav and Dr Iyengar — was constituted in early 2003. The team was asked to work on the project after hours and to come up with a new concept of personal transportation. They were provided *carte blanche* and a clean sheet.

Drawing Board to Driveway

The team tried various seating configurations, various body types, engine positions, different types of prime movers like conventional engines, electric vehicles, *et al.* "We did a detailed comparison of the pros and cons of each of these configurations, so that we could narrow down the potential options," explains Jai Bolar. In early 2003, there was a significant amount of buzz in the media after the Geneva Motor Show when Ratan Tata announced that Tata Motors was working on a 'one-lakh-rupee car'. That's when the team had to screen all the options using this new price filter. Considering that a three-wheeled auto-rickshaw sold at Rs 1.35 lakh, the team tried to imagine the lowest specifications for a car.

One of the early ideas it came up with was whether it could be a quadricycle. Though quadricycles had been in vogue in France since the 1980s, it was only in the 1990s that the European Union's regulatory bodies defined the rules that created a new genre of vehicles with this nomenclature. Essentially, four-wheeled motorcycles, quadricycles do not follow the same regulatory requirements as cars, most importantly in the aspect of crash and emission requirements. The standards applicable to quadricycles are somewhat similar to those for motorised two-wheelers.

In fact, there are two types of quadricycles: light and heavy. The maximum allowable kerb weight of the lighter ones is 350kg and the regular quadricycle, 400kg. This serves to regulate their use to essentially within cities and semi-urban/semi-rural areas, with a universal ban on their use on highways. As quadricycles could not be driven on highways and expressways, given their maximum speed is rather modest, the thinking was that safety requirements did not need to be as strict as for 'normal' cars.

Thus, a quadricycle by definition was not much more than two motorcycles joined at the hip; and so the thinking within the CET at Tata Motors was that a four-to-five-seat quadricycle could be a vehicle that would address the brief. The pricing of a four-wheeled motorcycle could easily be between that of a two-wheeler and the lowest priced car in India then — the Maruti 800. But the 800 had been in production for 16 years by then, thus costs were obviously amortized to the point where it was one of the cheapest in the world. To significantly undercut Maruti's bestseller seemed a rather difficult objective.

To be cheaper than the 800, something had to go, something had to be less; so, would a door-less quadricycle with a minimalist roof be the answer? Could vehicles like the Ford Think NEV or the DaimlerChrysler-backed GEM E825 be the answer, wondered Dr Sumantran and the CET? The Ford Think and the GEM E825 were two of several American NEV (Neighborhood Electric Vehicle) models, otherwise known as an LSV (Low Speed Vehicle), a category created in the USA in 1998 by the National Highway Traffic Safety Administration.

Similar to the French and European quadricycles in many ways, the LSVs were limited to a top speed of 25mph (40km/h) and were required to have basic safety equipment including headlamps, tail lamps, stop lamps, turn signals, mirrors, seatbelts, side reflectors, a parking brake and a safety-glass windshield. And like the European quadricycles, the LSVs too, were exempt from crash testing.

The Tata Motor team designed and made a rough prototype (a mule, in automotive parlance), of a quadricycle. To power the mule it used a 150cc air-cooled engine from a Bajaj motorcycle. The team quickly realised the quadricycle idea was not going to work. For one, team members felt that customers, especially in India, were not going to accept a vehicle that did not seem like a proper car, even if it was cheaper than the cheapest car. An apology of a car just would not do. For another, there was the not-so-small matter concerning legislation in India then.

For passenger cars, two-wheelers and commercial vehicles, especially for emission levels, Indian norms were quite similar to those in Europe; India was just a step or two behind in generational terms. And if and when crash tests and other safety norms came into vogue, India followed European norms. Yet, when the issue came up about introducing European norms for quadricycles in India (with automakers like Tata Motor and TVS Ltd supporting the idea), several other two-wheeler makers and car-makers lobbied strongly against it.

The Tata Motors logic was that quadricycles, with their four wheels, were far safer than two-wheeled motorcycles and scooters, which had traditionally been significant contributors to India's infamous road death score, the need for additional automotive-style safety standards were not necessary. "When I argued for quadricycle norms within SIAM (Society of Indian Automobile Manufacturers), surprisingly, the biggest support came from Venu Srinivasan of TVS Motors," said Dr Sumantran. "But the strongest resistance came from Jagdish Khattar of Maruti Udyog and Rahul Bajaj of Bajaj Auto. In fairness, I must say, Mr. Bajaj was very frank and transparent. He clearly stated that the proposal would be adverse to his company's interests at the time."

Clearly, good old monopolistic and protectionist tendencies once again reared its ugly head and subverted what could have been (and still could be), a good thing for the people of India... Ironically, Rajiv Bajaj, Rahul Bajaj's son and CEO and Managing Director of Bajaj Auto since 2005, had been pleading for quadricycle norms. The Bajaj answer (in terms of price) to the Nano, the quadricycle RE60, was unveiled in January 2012. In May 2013, the Government announced quadricycles would be allowed on Indian roads but would be restricted to commercial vehicles, as a substitute for three-wheeled auto-rickshaws.

Designing a 'proper people's car'

The quadricycle idea was abandoned by Tata Motors in 2004 and the decision was taken to develop a 'proper car'. Now, what is a 'proper car'? Can a car with just three doors be considered proper? Designs of this type had been in vogue in developed markets. Would people insist on having the option of air-conditioning? What was the minimum and maximum speed that would be acceptable? What would people want regarding fuel efficiency?

The truth was that the team at Tata Motors had no benchmark to go by. Yes, the Maruti 800 was there, and could be referenced, but it was already two decades out of date, and the game had moved on. Or, had it? Dr Sumantran, well aware of trends and technologies across the globe, had been quite taken by a concept that specialist French engineering, manufacturing and car-making firm, Matra, had shown at the Detroit Motor Show in early 2003 — the Matra P75. Just 3.54 metres long, the Matra P75 was one of Michelin's Challenge Design exhibits from France, a design that showcased the very practical 'one-box cubic' shape that had become a French obsession from the early 1980s.

Dr V Sumantran, Executive Director at Tata Motors from 2001 to September 2005, with German Formula One star, Sebastian Vettel, at the Indian Grand Prix of 2011. Courtesy Dr V Sumantran

Matra had been at the vanguard of the minivan or MPV (multi purpose vehicle) movement as defined by the path-breaking Espace, a seven-seat people-mover not unlike a Toyota Innova, that Matra had been making for Renault since 1984. It had also revisited the one-box idea several times, starting with the Matra P55 concept from 1993.

In the P55, Matra had showcased a similar concept with seating for four, but in a back-to-back arrangement, with the rear bench facing rearwards. The P75 had conventional seating for four, all facing forward, but with access to the rear compartment via symmetrical doors moving along the sides of the body. The frame consisted of two stainless steel plates sandwiching a layer of composites. An electric car concept, the P75 had its batteries and electrical traction system fitted under the floor. Very innovative, very cutting edge, but just too expensive for what Tata Motors was planning to develop.

The innovative Matra P75. Courtesy Matra

So, the process had to be iterative. "Starting from the very basic and using first principals was the only solution," explained Dr Sumantran. Yet the shape of the Matra P75 remained in Dr Sumantran's mind and the team developing the Nano.

The first review with Ratan Tata was in early 2003, when the team members presented the results of their study. This was before he had made the announcement in the media about the price of the car. "We had worked on the project over a few weeks after office hours and didn't think we were in any state of readiness to present to the Chairman of the group," confesses Jai Bolar. Anxiously, they put together the slides, trying to build a story; and travelled to Tata Motors headquarters at Bombay House. "We were thrilled and excited by the prospect of meeting Mr Tata, but at the same time nervous since we didn't know what to expect. When Mr Tata entered the room, there were just the four of us, led by Dr Sumantran, to make the presentation."

The first thing Ratan Tata did was to make them comfortable, saying he was not there as Chairman of the company but a team member who wanted to be part of the project. So it ended by being more a discussion than a presentation. Mr. Tata heard the team and came up with ideas. "We all left the room beaming with happiness and thrilled to be working with Mr Tata," remembers Bolar.

The first buck to evaluate packaging. Courtesy Dr V Sumantran

The team began to grow and a dedicated project management function was established. Fom December 2004, weekly reviews with the Chairman's office became a regular feature, as Mr Tata got more involved with his pet project. On 29 July 2005, Ravi Kant, one of the Executive Directors till then, was appointed Managing Director of Tata Motors. One of his first decisions was to get engineer Girish Wagh to take over the small car project. Wagh had impressed both Ravi Kant and Ratan Tata with his handling of the Tata Ace project for a small diesel-engine pick-up, which had been launched a few months earlier and was already setting the sales charts afire. Ravi Kant made it clear to Wagh at the outset that the small car project was top priority for the company and that Wagh was going to report directly to him, indicating that the progress on the project would be frequently reviewed by Ratan Tata himself.

Ravi Kant took over after Dr Sumantran left. © Makarand Baokar

The first step for Wagh was to acquaint himself with the product development situation and R R Akarte, Head of Tata's Engineering Research Centre, briefed him. Wagh realised that it was important to finalise the concept of the various mechanical aggregates and the body along with the styling. At the same time, it was as important to decide on a manufacturing strategy and establish the supply chain. For the latter, he turned to B B Parekh, the Tata Motors Head of Strategic Sourcing. Also, he put in some thinking on building the organisation and team that would make the car happen. As the team evolved, so did product development.

Narendra Kumar Jain, Deputy General Manager (Engines), who had already designed the Tata Motors first gasoline engine, had been scouring the world for an appropriate small engine Tata Motors could buy out and use to power the car. But none seemed to be available. "We did take a look at the Russian small car, the VAZ Oka, but it was too crude," explained Dr Sumantran. The Oka was a small Russian-designed 'peoples' car' from 1988, that featured a compact 750cc two-cylinder engine.

Dr Sumantran also considered one of the last two small twin-cylindered Japanese engines in production — the EH series from Honda — which the Japanese car-maker had been using for its Acty light truck and the Today car from 1977 to May 1988 (the other being the 546cc 2G23 engine used in the fifth-generation Mitsubishi Minica from 1984 to 1987). Based on one bank of cylinders from the Gold Wing GL1000 motorcycle, Honda's 545cc engine used to develop as much as 31bhp then. "If the Honda Today was powerful enough for Japanese customers of the 1980s, perhaps it would be good enough for Indians today," surmised Sumantran. So he had discussions with two of Honda's two-wheeler collaborators, Hero Honda and Kinetic Engineering. But it was obvious that getting the Japanese giant to resuscitate a mothballed engine was going to be too complicated. "We also met Austrian and American engine manufacturers Rotax and Briggs & Stratton, but Rotax was too expensive and the latter's powerpacks were not very refined," explained Sumantran.

Hi-tech versus low-cost

The decision was taken to develop the engine in-house. Obviously, a 150cc air-cooled single, barely big enough to propel a motorcycle, was deemed insufficient. It had to be bigger, at least a twin-cylinder — the bigger a single-cylinder engine, the more it 'rocks' and vibrates, making its application in cars unattractive. Clearly, a single-cylinder-engine car would be perceived as unrefined.

So a water-cooled twin of 554cc was first designed. It was also decided that it would be an in-line twin — the cylinders placed one after the other — for reasons of cost and simplicity. "Costs, drivability and fuel efficiency were the three paramount objectives when developing the engine," said Dr Sumantran, adding, "we had to constantly keep in mind our 'out-of-the-door' target price of Rs 75,000, and so the engine couldn't be more than Rs 25,000 in cost!"

It was also decided that the engine would be vertical, as against lying flat (incidentally, the Citroen 2CV engine and that of the Volkswagen Beetle's lie flat), to save on length and space, to keep the road encumbrance (the footprint of the car), at a minimum. A running mule — more car-like than the quadricycle — was readied with this engine in December 2005. In the meantime, by September 2005, Dr Sumantran had moved on.

Ratan Tata drove the mule. He was not impressed. He felt that customer expectations — in terms of drivability — would not be met by this 20 horsepower engine. It was back-to-the-drawing-board for the team.

Bore and stroke was increased and the swept volume went up to 586cc, with max power going up to 27bhp. With that, the drivability of the mule improved and was closer to the perceived (or assumed?) requirement. But to be competitive it was felt that a little more power was needed. Bore and stroke went up once again and the engine size grew to 624cc, with max power reaching 34.5bhp (35PS). Using simulation, the team came to the conclusion that this size was optimum for the targeted power-to-weight ratio, for future emission norms and handling an air-conditioning system.

Now why did the team design a twin and not a triple-cylinder engine as in the Maruti 800, or as can be seen in many Japanese small cars (most have 660cc triples)? A twin is cheaper than a triple, for one, plus it is more fuel-efficient. With a triple you can get more power, true, but then power was not a priority, whereas fuel efficiency was. Also, when developing the twin-cylinder engine, the options for arranging the cylinders were many: the team could have designed a horizontally-opposed engine, or a V-twin, yet it chose to go with an in-line twin. Why? A horizontally-opposed twin would have been lower, with a lower centre of gravity, ensuring much smoother movement as the rocking motion of the two cylinders firing 'against' each other minimize and negate vibrations. A V-twin could also be quite smooth (depending on the angle between the two) and it too would have been lower, but both these engine arrangements would have cost significantly more, and required greater manufacturing complexity. Remember, keeping costs down was top priority.

To address the problem of roughness, engineer Narendra Jain designed a balancer shaft, an eccentric weighted shaft that offset and smoothens vibrations. In the process, the Nano's twin became the world's first twin-cylinder engine with a balancer shaft for a car — a patentable solution! Truth be told, balancer shafts are not new. In fact, they hark back to the very dawn of the motor car with British engineer Frederick Lanchester, the first to invent and patent this idea way back in 1904. The unpredictable twists and turns of the automotive industry being what they are, one can speculate that Frederick Lanchester, in his wildest dreams, would never have imagined that the car marque he founded with his brothers George and Frank, in

December 1899, the Lanchester Motor Company (also credited with being the first purely English car-maker), would one day belong to an Indian car-truck manufacturer called Tata Motors!

When Tata Motors bought Jaguar Land Rover from Ford, it also inherited several other brands, including Rover (which Ford had acquired from BMW when the latter sold MG Rover to the Chinese car-maker Nanjing Automobile Group), Daimler and Lanchester, the last having been taken over by BSA in 1931, following the 'Great Depression', when Lanchester went broke, along with Daimler. BSA, in turn, sold Daimler and Lanchester to Jaguar in 1960.

THE LANCHESTER CONNECTION COMES FULL CIRCLE

A no less interesting footnote is the possibility that the Lanchester connection to India is perhaps older than the 'since-2008' ownership of the brand by Tata Motors. One of Frederick Lanchester's closest friends was a certain Indian prince, Kumar Shri Ranjitsinhji Vibhaji Jadeja, otherwise better known as Ranji. Yes, the same Ranji of Ranji Trophy fame. Ranji became the Maharaja of Nawanagar, a princely state in Gujarat, in 1907, but before that he was already famous as one of the world's finest cricketers.

After attending Trinity College, Cambridge, Ranji played for the cricket team of the county of Sussex in the UK and served as the team captain. In 1896, Ranji joined the English cricket team, becoming the first Indian to play top level international cricket; he went on to play for England till 1902. Widely regarded as one of the greatest players of all time, Ranji's playing was described by Sir John Frederick Neville Cardus, an English writer and critic, best known for his writing on music and cricket, as "the Midsummer Night's Dream of cricket".

Playing and staying in England at long stretches, Ranji befriended the brothers Frederick, George and Frank Lanchester. Ranji bought his first Lanchester, a 12hp twin-cylinder model in 1904, with which he competed in the Poona Trials in India. From that year onwards whenever a new model was introduced by Lanchester, at least one was shipped to India to add to Ranji's Lanchester collection. It's not that the Maharaja of Nawanagar bought only Lanchesters — he had a fine collection of cars from other marques too, brands like Bentley, Bugatti, Chevrolet, Lagonda and Rolls-Royce — but his collection of Britain's Lanchesters was the most extensive, with at the least 42 Lanchesters finding their way to the Nawanagar stable till 1940 (by then Ranji was dead, but his nephew who became the maharaja, carried on with the tradition too).

Other than being a good friend of Frederick Lanchester who spent many winters with Ranji in India, it's believed that the maharaja may have also had some shares in the car-maker too, though there's no clear evidence to support this hypothesis, but it perhaps explains his desire to be a good client by example. So it is quite possible that Tata may not have been the first Indian to have had a stake in Lanchester....

The use of the balancer shaft in modern motoring dates back to 1975, with Mitsubishi pioneering its use for its four-cylinder engine. Mitsubishi went on to license its patent to Fiat, Saab and even Porsche. Though most of the industry uses it now — with parallel twin and single-cylinder engines in motorcycles also using balancer shafts — its use in an in-line two-cylinder car engine was pioneered with the Nano, thus the opportunity of cornering a patent, along with nine more for the engine and 24 more for other components.

In the meantime, development work on the body was going on. Once the quadricycle idea had been abandoned and the decision taken to make a 'proper car', packaging and engine location became the crucial next steps. Two or three-doors that allowed for a single side (left) entry to the rear, was seen as something customers would reckon to be an incomplete car. So the decision was taken to go with four doors and a 'proper' body.

Ratan Tata was also keen on pushing the technology envelope, so the idea of using composites or plastics was examined. The possibility of designing a body with a steel space-frame clad with plastic panels — the design of most French quadricycles — was considered seriously but costs and manufacturability issues, along with Tata Motors' inexperience at alternative composite technology, put paid to such ideas. Instead, the decision was made to go back to old-fashioned steel, with a difference: the body construction, though a monocoque (as per current practices), featured two longeron-like strengthening sections running longitudinally. These are cross-braced with a pair of cross-members running transversely, the front one a few centimeters aft of the front wheels, the rear ahead of the rear wheels. Tata Motors called it a hybrid-monocoque and it was done to save weight and complexity. With a stiffer floor, the sheet steel body used the minimum of double skinning to keep weight down and minimise tooling investment. To improve body rigidity, a transversely located rod between the B-pillars acted as another chassis 'member', other than being used as a mounting point for the driver and front passenger seat runners.

Similarly, for the styling, an iterative process ensued. The initial doodles were done by Tata's own Nikhil Jadhav. Then the job was given to Italian design house, I.De.A Institute, the same outfit that designed the Indica. A team, led by Justyn Norek, began rendering some evocative but simple mono-volume shapes for the one-lakh-rupee car.

But before that, packaging issues had to be decided — specifically whether the engine would be located at the front, at the rear, or elsewhere. Most small cars in production today have their engines at the front with drive going to the front wheels. But in packaging terms, this is not necessarily the best solution. Instead, placing a compact powerpack aft and under the rear seat makes a lot of sense if you want to make a car that is small and space-efficient. Starting with the Smart (which was launched in 1998 by MCC, a subsidiary of Daimler-Benz), and followed by the Mitsubishi i (launched in 2006), increasingly more car-makers have been exploring the idea of putting a compact engine at the rear of their small car range.

Other than packaging, the other advantages are in the mechanical simplicity that you get, as you can say goodbye to CV (constant velocity) joints, needed to allow a drive shaft to transmit power as wheels turn and rotate at different speeds. Also, engine noise insulation becomes more manageable and engine heat is less of an issue as the seat backrests act as insulators. With less weight on the front wheels, power steering is, in theory, not necessary as the steering becomes lighter.

There could be an issue with vehicle dynamics though: rear-engine cars like the Volkswagen Beetle and Renault 4CV, have 'tricky' handling, whereas front-engine front wheel-drive cars are better able to negate mistakes and are known to be more forgiving. But then, the Beetle and Renault have their engines 'overhung' at the rear, located longitudinally aft of the axle line, with the drive going forward to a gearbox and transaxle that transmits power to the rear wheels. With the engine located far out in the rear, these cars have a high polar moment of inertia and when cornering at speed, the tail tends to 'step-out', causing the car to spin.

With the car that Tata was working on, this seemed to be less of an issue as the engine was not particularly powerful, and neither was it 'overhung'. Narendra Jain's design was for a compact, twin-cylinder layout that could be placed transversely just ahead of the rear axle line — slightly canted over and off-centre to the right — with drive going sideways to the gearbox-transaxle unit to the left, from which the unequal length drive shafts took the power to the rear wheels. To make the engine bay as compact as possible, the radiator was placed on the right, aft of the engine, and the exhaust manifold snaked to the left, curling around to a centrally located exhaust.

In the meantime, as the styling iterations continued, with the decision to go with a conventional steel-bodied car, an alpha prototype (reflecting the definitive design proposal), was made, and which Ratan Tata personally tried out in July 2006. He was not happy with driver ingress and egress. Neither did he care for A-pillar visibility and control access. So it was back to the packaging drawing board.

On the early design proposals by I.De.A Institute. Courtesy Dr V Sumantran

Rear view proposal by I.De.A Institute. Courtesy Dr V Sumantran

The vehicle length went up by 10cm, the wheelbase by 5cm and the aperture for the front pair of doors was increased by 3.5cm. Another redesign followed. As the mechanicals and the exterior design were getting frozen, work on the interiors was progressing rapidly. With space, comfort and functionality as priority, other than the reality-check of keeping costs down, work on designing the interiors began. Locating the instrument cluster at the centre was done so that everyone could look at it, other than the fact that, with such a dashboard design, the same tooling could be used for both right and left-hand drive. Similarly, the two individual front seats were designed to be identical so that the same tooling could be used. The seat frame remained a simple metal structure, with the headrest integrated with the backrest into a single piece to save on cost (but, of course!).

As design and development of the car was going on, work had also begun on the manufacturing side of the activity and, by November 2005, the exercise of locating the factory site had begun in real earnest. By early 2006, a full-fledged manufacturing team had been formed. And on May 17, 2006, the selection of Singur in West Bengal as the factory site was announced. Ratan Tata was very keen that his small car project would also bring back industry to West Bengal, along with jobs.

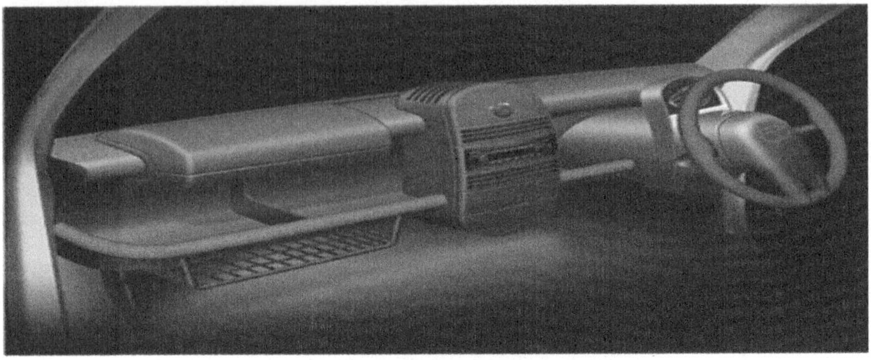

One of the proposals by I.De.A of the fascia. (courtesy Dr V Sumantran)

Simultaneously with design refinement and engineering of the car, vendors were getting involved in the process too, not just as potential low-cost suppliers, but also in the development and design of some of the components. German component-maker Bosch, was roped in and got involved in developing an inexpensive fuel injection system along with an engine management system that ensured fuel efficiency and emission standards were met. Sona Koyo, which specialised in steering systems, developed a hollow steering wheel column to reduce weight and cost. Tata Yazaki designed a new fuse box based on the bus bar technology, helping in reducing 21 circuits without compromising functionality. At every component development exercise, cost reduction was the prime concern: examples included the inner rear view mirror which also incorporates the cabin lighting system, the single wiper with built-in washer nozzles, the example of integrating a honeycomb grille into the bezel of the dashboard vents, and so on.

But before the suppliers could get going on readying the tools and dies for production, Tata needed to sign off on the final design of the car. And so a beta prototype (with a more definitive design) was finally ready by mid-2007 and the process of signing off had begun. At the same time, the installing of equipment at the Singur factory started. Into the last quarter of 2007, the design and engineering specs were frozen, pre-production prototypes readied and testing had begun. Three of these pre-production prototypes became the white, red and yellow cars on the Tata stand at the unveiling of the Nano on 10 January 2008, at Delhi's Auto Expo. And two more pre-production prototypes became the Geneva Motor Show cars.

But soon thereafter, by late 2008, things started to go a little pear shaped. The political impasse at the Tata Nano's factory site of Singur in West Bengal, forced the car-maker to move out; the decision was taken to relocate the factory to Sanand, in Gujarat, with interim production at Tata's light truck factory at Pantnagar in Uttarakhand, This inevitably delayed the launch of the Nano.

The Tata Nano, after its unveiling at the 2008 edition of the Geneva motor show, had every car-maker executive more than curious.
© Gautam Sen

Even if the wait for the car was getting a little too long, Tata still managed to pull off another media coup when the Tata Nano was finally launched on 23 March 2009, in Mumbai. It was another razzamatazz affair and once again an event that had both the Indian and international media laying out column inches on the Nano. The official launch of the l'il car was not lacking in pomp and ceremony either. Symbolically, Tata decided to have the press launch at the Taj Mahal Hotel, Mumbai, which is owned by the Tata Group and was a target of terrorist attacks in November 2008. After the elegance of the Taj Mahal Hotel's ballroom, the next stop in the Nano's Mumbai media launch was a ten-minute drive away, to an even more elaborate outdoor press conference held on the grounds of the Parsi Gymkhana, facing Mumbai's famous Marine Drive and the Arabian Sea. Journalists, executives, Tata dealers and other invitees, watched as lights flashed, music blared, and three cars slowly rolled onto the stage. The Tata Nano had arrived.

Starting from 9 April 2009, Tata Motors began taking advance bookings for their 'little wonder' and by the time bookings had closed on 25 April 2009, exactly 206,703 Nanos had been booked by eager punters.

Driving the Nano

On 17 July 2009, Ashok Raghunath Vichare of Mumbai, became the first owner of a Tata Nano. Ratan Tata, at a Tata dealer outlet in south Mumbai, handed over the keys. For Vichare, it was not just his 15-minutes-of-fame moment, but the moment he became the proud owner of his very first car. Like Harpal Singh of Maruti 800 fame, Vichare was not just acquiring a slice of history, but enacting it, a little over a quarter century after India's first proper go at a car-for-the-people rolled off the assembly lines in Gurgaon.

So, did the car that Vichare and thousands of others who started taking delivery from 17 July 2009, meet expectations? Did the Nano drive like a 'proper car'? Was the design likeable? Did the car appeal to everyone?

On the looks front, the car has been controversial. Many feel it is just too small, upright and ungainly. Others believe the Nano is cute, even if has a rollerskates-like look about it, especially the dinky 12-inch wheels that look too small for the 1.6-metre high car.

The design is very much the 'monovolume' type, or what the French term 'monospace', which, for the Nano, is perhaps appropriate. A lot like Maruti Suzuki's Zen Estilo, the Nano's profile is a clean curve that comes down all the way from the top of the greenhouse to the leading edge of the front bumper, with the line defined by the A-pillar curving down inwards to meet its symmetrical pair in a sort of a V-shaped smile, that is supposed to exemplify the Tata car design identity and the 'face' of all Tata cars. A pair of smart, clear-lens, trapezoidal headlamp units (incorporating the indicators), sits above the smile. You will wonder about the effectiveness of the single wiper (which, incidentally, incorporates the washer jets within it), but given the shape of the windscreen, which is almost as high as it is wide, the wiper is very effective. Incidentally, Mercedes-Benzes, from two generations ago, had a single wiper system, as did the Fiat Uno.

The rear reflects similar clean lines with the typical Tata vertically-stacked lamps — a trend started by the first generation Indica — providing some visual relief. Of course, locating it high also serves the twin purposes of protection and better visibility. A discrete grille takes up most of the bottom of the rear bumper, providing the vents for extraction of hot air from the engine compartment.

The rear windscreen is bonded to the rear section as it contributes to the necessary stiffness and torsional rigidity of the body. At the time of the launch, Ratan Tata himself admitted that this aspect needed a redesign as there was popular demand for an opening rear hatch for access to the boot area. Behind the rear bench there is some luggage space — all of 80 litres and good for one biggish suitcase vertically, and about four sets of toothbrushes. But to get to those toothbrushes one needs to fold the rear bench down, with access through either of the rear doors. Tata Motors promised to address this issue by developing a simple glass-only tailgate, much like the first generation Maruti 800 used to have.

If there are just two people aboard, the rear bench can be folded flat, in which case the luggage space expands to a much more impressive 500 litres. Access to the engine is in the same way, incidentally — by folding the rear seats, one gets to the 'floor' of the boot area, which also serves as the engine cover. Then six flynuts have to be undone before the panel can be removed and the engine accessed. Not one bit convenient.

From the side, the Nano has the slightly 'wedgy' egg look — very clean with minimal ornamentation. Other than the neat curve of the greenhouse starting from a midway point of the rear lamp cluster, sweeping all the way to the leading edge of the A-pillar and beyond to the front bumper and the straight rising beltline, the flanks of the Nano have one neat little design touch: an upward rising swage line that kisses the top end of the air intake vent along the trailing edge of the rear door, giving the car a slash of sportiness. Another noteworthy visual aspect is the ribbed roof structure. But it is not a design element. In fact, it is another detail feature that contributes to the roof's rigidity, which in turn, allows for the use of sheet metal of a thinner gauge. The ribs, other than providing the necessary strength, are a nice styling touch.

Despite a length of just 3.1 metres — almost 24cm shorter than the tiny Maruti 800 (68cm shorter than a Citron 2CV and 95cm shorter than a VW Beetle) — the Tata Nano does not look all that small. It is higher than the 800 by almost 25cm, and wider too, by more than five centimeters. By making it higher and wider, interior volume is significantly more, aided by the fact that the engine is located at the rear. This becomes more obvious when you consider that at 2.23 metres, the Nano's wheelbase is 5.5cm *longer* than that of the significantly longer Maruti 800. As Tata Motors claims, the Nano is 20 per cent more spacious than a Maruti 800, yet the road encumbrance (the car's footprint) is eight per cent less. And that is immediately obvious when one gets into the car. The sense of space is much more than in the Maruti 800; in fact, it is more like the Maruti Alto. For another, one sits much higher and the view through the deep, wide windscreen is superb, with visibility better than with many of the other small cars in the marketplace and not dissimilar to the Hyundai Santro.

How has that been managed? It has been possible due to the higher seating, with the H-point, the location of the hip that is, higher than for most small cars in India, which, incidentally, also makes the business of getting in and out (ingress and egress), that much easier (and more elegant) in the Nano.

Once inside, what strikes one is that the car's interiors are Spartan. The dash is a simple one-piece moulding that houses the instrumentation — essentially a speedometer, along with digital readouts for fuel level and temperature — which is high up at the centre, quite in the vision of the driver, without obstructing it, allowing the same moulding to work for both left and right-hand-drive versions. On both sides of the instrumentation, there are hollowed-out shelves that can take odds and ends. When launched, there was no glovebox or lidded compartments, an addition made in the latest iteration, in 2012.

Adorning the dashboard are four face-level vents for fresh air and/or air-conditioning for the higher trim variants; for the base version there were no vents at all. The centre console of the dash has the airconditioning controls and instrumentation, and a cutout for an audio system if the car buyer wishes to opt for one. For the base version, the steering wheel is a two-spoke affair; the higher end versions get sportier, more substantial three-spoke wheels.

In the Nano, one sits high, a little van-like in the stance, but the steering-wheel angle is very much car-like, with enough space in the footwell area for a pair of shoe size 12s. In fact, leg space is impressive for both the driver and the passenger in front, as well as for passengers at the rear. Ergonomically (relation of the driver's seating position to the steering wheel and pedals) the car is good. But one has to remember that for the base version, only the driver side seat is adjustable and the front passenger's is fixed. For the mid-level and top-end versions, both the driver's and front passenger seats are adjustable.

What one may not notice, though, is that the driver has slightly more space available than the passenger next to him. That is so because the driver's seat is located slightly to the left of where it should have been, whereas the front passenger's seat has got 'pushed' further to the left. This has been done for two reasons: to keep tooling costs down, the two seats are identical with the lever for seat backrest adjustment for both the seats located on the right side, which has necessitated the driver seat to be located a little left of where it ought to have been.

The other reason is that the Nano was originally conceived as an automatic, which would have meant that only two pedals (accelerator and the brake), would be on the floor. Midway through the development of the vehicle, the decision was taken to launch the car with a manual gearbox, necessitating the addition of another pedal for the clutch, which was located further left of the brake pedal, and to get the ergonomics right, the seat had to be moved slightly to the left.

Either way, the ergonomics is spot on, and there is more than sufficient space for two well-built adults up front. Similarly, two extremely well-built adults can sit comfortably side-by-side at the rear, and even three slim-and-trim types can squeeze in. That has been possible by keeping the 'tumblehome' of the Nano much less pronounced and the sides rather vertical. What's a tumblehome, you may ask? The tumblehome is the inward slope of the 'greenhouse' above the beltline. Today, more and more cars have pronounced (more curved), tumblehomes in order to reduce aerodynamic drag. But as the Nano is not a particularly quick car and aerodynamic drag is less of an issue, the need for the tumblehome to be pronounced was less. So the car was designed to be rather vertical along the flanks. By being so vertical, the Nano manages to free up more shoulder space.

The other advantage of the 'verticality' was that the curve of the winding side windows could be kept at the very minimum, with just a slight single curvature, which allowed for easier winding mechanisms, and narrower doors. All of which contributed to the business of reducing costs and investments in tooling. Incidentally, the quarter windows at the front are flat, making them that much cheaper. But for the sake of style, the front and rear windscreens are the more expensive compound curvature types.

The other aspect that is striking, other than the sense of space, is the sense of airiness. With six light windows on the sides, the deep windscreen up front and a decently large glass area at the rear, the visibility out of the Nano is very good. The only 'dead' angle is thanks to the slightly thicker pillar at the rear quarters, but then it contributes to a feeling of strength.

Visually, there are differences between the base and intermediate versions and the higher end trim level: the base version is distinguishable by its unpainted black bumpers, black driver side door mirror and steel wheels with grease caps; the CX too comes with black bumpers and black mirrors, but the wheels get centre section plastic caps. Only the LX gets body-coloured bumpers and door mirrors, plus smart plastic wheel caps. The base version is the bare-boned one — no air-conditioning, no features like the body-coloured bumpers, etc.. The CX gets air-conditioning (along with a tinted windscreen) and seats with fabric inserts; the LX gets air-conditioning, tinted windscreen, plus electrically-powered windows (for the front two windows only) and central locking, cigarette lighter socket, fog lamps, an extra stalk on the steering column, full fabric seats and door pockets.

There is also another less obvious reason why the engine has been located at the rear: for safety reasons. Contrary to popular belief, it is more difficult to engineer a car's safety when the engine is at the front, as in the case of a head-on or front offset crash, when the engine tends to push back into the passenger area and can do considerable damage. Whereas, when there is no engine at the front, the area up front can be used to design a 'crumple zone' that can absorb or offset a frontal impact better.

SAFETY ISSUES

Illustrating the point on increased safety with engines places in the rear, is the Smart. The little car received a four-star rating for its crash tests, which is very good, as compared to the front-engine Hyundai Atoz (the Santro as it was called in Europe), which managed just three stars when crash tested by the independent agency, EuroNCAP (European New Car Assessment Programme). Incidentally, the engineers at Tata Motors say that the Nano meets not only Indian, but also current European safety norms, as it is.

The fact is that European safety norms is one thing and EuroNCAP is another thing, and that needs a bit of clarifying. EuroNCAP, which was initiated by FIA (Federation Internationale de l'Automobile), is an independent testing system by which motoring consumers – both drivers and the automotive industry — is provided with a realistic assessment and relative safety performances of some of the more popular cars sold in Europe. Using a system of star rating from one to five, EuroNCAP, independently of the European automotive certifying agencies, have been rating cars.

So even if a car has a very poor one star rating as per EuroNCAP, it is quite possible that the car may have passed European crash test certification. The Nano, as it is for India, should pass the minimum requirements that will allow it be certified for sale in Europe, but may not be able to garner a lot of stars as per EuroNCAP. And to sell in Europe the number of EuroNCAP stars is more important from the consumer point of view — thus the need to re-engineer the car for Europe and North America.

The other technical features of the Nano are a study in minimalism: the car has drum brakes at all four wheels, no power steering and just a four-speed gearbox. The first two also result from engine location. As the engine is at the rear, and not at the front, the weight bias is also towards the rear, with 60 per cent of the vehicle's kerb weight on the rear wheels and just 40 per cent on the front. And given that the Nano is a lightweight 600kgs (for the base version that is), the weight on the front is so little (just 240kg when unladen) that the steering, in theory, does not need to be powered, and neither do the brakes need to be discs (which, though more efficient, are more expensive). At least, that is what the tech boffins from Tata Motors have been justifying. Incidentally, the higher trim versions — which are 615kgs heavy in CX trim and 635kgs for the LX — the brakes have servo assistance, which makes braking more effortless than with the base version.

The four-speed gearbox is another case of cost cutting, as it is cheaper than a five-speeder, and the CVT (continuously variable transmission), which is an efficient automatic like those used in scooters and which Tata were planning on initially. Also, given the max speed of the Nano has been restricted to 106kph (for reasons which will be explained later), the engineers did not see the need for a more expensive five-speed gearbox.

The other rather unusual technicality has to do with the tyres: the rear ones are wider than the ones in front, other than being set wider too. Why has that been done? With a rear engine position and rearward weight bias, a car's handling could be of the kind where predictable understeer switches to sudden oversteer. In layman terms, it is a situation somewhat like this: you are taking a turn, and the car seems reluctant to do your bidding, so you try to force the issue with greater steering inputs, and that's when the tail suddenly slides out, resulting in a dangerous spin. Great for action movies but not for everyday driving.

To address this, the techies at Tata Motors equipped the Nano with rear wheels that are wider than the front — 155mm wide versus 135mm, all on 12-inch wheels — and located a little wider apart (the rear track is markedly wider than the front), so that the rear is able to retain grip better, whereas the front can concentrate on the business of steering the car. For rear-engine cars, this is not unusual: the Smart's rear tyres are fatter and wider apart than the front wheels, and so is it with one of the greatest sports cars of all time — the Porsche 911.

The tyres, incidentally, are tubeless radials, that are dynamically superior to conventional tyres. Tata Motors used tubeless tyres to reduce weight — to the extent of two kilograms across the four tyres — which may not seem much, but contributes significantly to reducing 'unsprung' weight, an attribute that improves handling and vehicle dynamics.

Like other rear-engine cars, the spare wheel is in front, under the tiny 'bonnet', and this spare wheel-and-tyre is of the smaller size, which can, when necessary, be used at the rear too, till the punctured tyre is repaired. The bonnet also provides access to the fuel filler cap, which is located at the front, though the fuel tank is under the floorboard of the rear passengers. All this has been done to offset the rearward weight bias. Other such measures include locating the battery under the driver's seat and the jack under the passenger's. Why did they not put the battery up-front even though there is space? It seems the extra length of the battery cable would have resulted in higher costs. That is also the reason why the power window switches are placed, most inconveniently, on the floor console instead of on the doors, since this would have meant using two separate wiring looms.

Keeping Indian conditions in mind, the ground clearance is a decent 18cm (a centimeter more than the Maruti 800), and the car has independent suspensions at both front and rear — traditional McPherson struts at the front and semi-trailing arms at the rear. To save on cost and complexity, there are no anti-roll bars. Brakes are drums on all four wheels and that may be surprising as even the Maruti 800 had disc brakes up-front when launched more than two-and a-half decades earlier. But as the overall weight is not high and the front is anyway lighter, the engineers did not see the need to equip the car with more expensive discs at either end.

Some of the out-of-the-box thinking includes the rubber air deflector at the rear, just below the radiator (there to deflect air from under the floor and divert it to the radiator, located at the rear, incidentally), enabling the use of a smaller and thus cheaper, cooling fan. The door handles on all four doors are identical; just one external lock has been provided on the driver's door. The wheels are secured with just three bolts, not four or five as on most other cars, as three is cheaper than four or five.

The Nano has been designed with the paradigm of eliminating costs: why have something that is not absolutely necessary? When it is possible to make things simpler, lighter, cheaper, why not do so? But then what is the car like to drive? Is it as compromised as some of its features?

The doors are big and they open wide. Step inside and the first thing you notice is the sense of space — the car does not feel small at all. Depending on how tall you are, you do not slide into the seats, but step in; as the seats are fairly high you sit high, with a commanding view out. Though the seats look very basic and rather thin and flat, they are comfortable. The driver's seat is adjustable for all versions; the front passenger seat only for the more expensive CX and the LX versions.

The steering wheel is small, though nicely thick-rimmed. The relationship between the seats, the steering wheel and the pedals is good. For the more expensive version, the steering wheel is a fairly sporty three-spoke affair, but the base version makes do with a simpler two-spoke steering wheel. The dashboard too, for the base version, is a plain sculpted slab that has the very basic instrumentation at the centre, and nothing else, other than the deep troughs running on both sides of the central instrument, useful for keeping odds and ends. The instrumentation consists of a speedometer only — the other gauges, which are electronic — show when the ignition key is switched on. And just one stalk on the right does the job of switching on the lamps, indicating and triggering the wiper for the base and the CX versions; the LX version has two stalks.

The single wiper with built-in washer nozzles, other than helping to save costs, works well, as the shape of the deep windscreen facilitates a wide sweep that clears most of the screen, leaving out a small triangle at the top left only, which is hardly an issue in terms of visibility (at least for the right-hand-drive version of the car).

It is a pity Tata Motors did not initially provide a door mirror on the passenger side, which is a safety essential, though hardly acknowledged as such in India, as the driver side door mirror, along with the inside rearview mirror, despite managing to cover most of the rearward vision, leaves a pronounced blind spot on the rear-left of the car. It was only in 2012 that Tata Motors decided to rectify this shortcoming by providing a passenger side door mirror for the CX and the LX versions. The base version still has one door mirror, on the driver side.

For the CX and LX versions, the dashboard is more elaborate, with four eyeball vents — two at the centre and two at the extremities — and a centre console incorporating the air-conditioning controls, and a cut-out for locating an audio system if you so wish. The air-conditioning controls are the rotating dial type and, in addition to temperature and blower speed control, it is possible to manage the direction of airflow too. In the LX version, the electric window switches are located just ahead and right of the gear lever, requiring quite a bend to get to it, with the plug for the lighter to their left.

Practically, other than a rather difficult access to the boot, the rear engine also heats up the boot area, with more than one wag describing it as a 'tandoor'!

Start up the Nano and you immediately know it is powered by a small engine placed somewhere behind you. The two-cylinder 624cc unit is an all-aluminium square unit with identical bore and stroke of 73.5mm. With two valves per cylinder and a single overhead camshaft (SOHC) actuating them, the petrol engine initially developed 35PS, but then a 2012 upgrade saw engine power rising to 38PS, which is about 37bhp. That translates to a specific output of 59bhp per litre and a power-to-weight ratio of 62bhp/tonne, neither of which is all that impressive. But then the maximum power has been kept lower than what was possible to make sure that fuel efficiency was good. The maximum torque is now 51Nm (up from 48Nm, the torque figure the Nano had at launch), again not a particularly impressive figure but adequate for the 600kg car.

Despite the balancer shaft technology used (mentioned in the last chapter as one of the patentable 'innovations' of the Nano), the powerpack is not very smooth or refined and sounds a bit like an auto-rickshaw. Though the engineers at Tata Motors have obviously worked at isolating the sound from the passenger compartment, they have not managed to isolate all the vibrations.

The gear lever has a smart design, with the top being rounded and the sides cut flat to allow you to use the thumb and middle finger grip with the index finger resting on top for easy shifting. The four-speed gearbox, placed transversely at the rear, has two push-pull cables operating the shifter mechanism. The shift quality is light, though not all that precise, though not too bad given that linkages for gearboxes located at the rear is always rather complicated.

Push down on the accelerator and the Nano gets going with all the grace it can muster; the low first and second gears, along with the ignition cut-out operating at a somewhat low 5600rpm, means you have to shift up quickly. The first gear is good for 30kph, the second gets you to 60 and the third takes you to 95kph. The fourth gear is for overdrive and the max speed is, in any case, governed at 106kph. As there is no tachometer, it also means you often end up revving it to the point where the engine management system starts cutting fuel supply. The focus was on setting up the engine and matching the gear ratios to get the maximum fuel efficiency along with low emissions, given the kind of drivability needed to make the most of low-speed driving that is typical in Indian urban conditions.

In that sense, the Nano does a good job: with a full load of four adults and the air-conditioning running, it is possible to merge with the traffic. Of course, without the air-conditioning, the Nano feels a mite more sprightly and is slightly quicker whilst accelerating. Tata Motors says that unlike other cars, the benchmark for the Nano was not accelerating from 0 to 100kph (as it is an urban vehicle), but from 0 to 60, which takes about eight seconds. To get to 100kph it does take a lifetime though... 30-odd seconds. Though that may not sound impressive, the fact is the Nano can cruise all day long between 80 and 90kph, without missing a beat, and, given that the fastest highway in India (the Ahmedabad-Baroda Expressway), is speed-limited to 100kph, the Nano is (just about), capable of doing highway duty too. Incidentally, the legal speed limit in the cities is 50kph and 60kph or 80kph on most Indian highways.

But it is in the nip and tuck of everyday traffic that the Nano is at its best. Thanks to the Nano's agility, its light handling and reasonably good drivability, it is possible to dart into gaps and weave in and out of traffic. The tight turning circle of just eight metres makes the business of taking sharp U-turns in our narrow streets a convenient twirling of the wheel. But parking the Nano is not the doddle you would imagine. Power steering, whilst parking, is sorely missed, despite the limited weight over the front wheels and the use of relatively skinny tyres.

On the move though, the rack and pinion steering is accurate and offers fairly decent feedback. But the tall body and narrow track tend to produce quite a bit of body roll. This is even more pronounced as the flattish seats, without any side bolsters, do not hold you as you slide all over.

Given the rearward weight bias, one would expect the handling of the car to be a little hairy. When pushed, rear-engine cars tend to oversteer on the limit, causing them to spin, and go backwards off the roads. But that does not seem to be an issue with the Nano. That bit of a wider rear tyre and a narrower front tyre (155/65 R12 versus 135/70 R12), plus the bit of placing the rear tyres wider apart than those at the front, have helped matters, while the decision to limit max speed to 106kph has taken care of the adventurous minority who may wish to explore the limits of the Nano's handling capabilities.

In third or fourth gears, which are taller gears in which speeds can be well over 60kph, you can throw the Nano around with impunity. It refuses to lose grip, despite the worrisome roll angles. At worst, the speed gets scrubbed away, never allowing you to explore the limits of adhesion. Anti-roll bars could have addressed this issue, but then it would have added to the cost — it was simpler to restrict the top speed.

The Nano uses drum brakes for all four corners, a little surprising when just about every car, including the two-and-a-half decade old Maruti 800, has disc brakes, at least in the front. Once again, the engineers at Tata Motors explain that given the weight on the front wheels is not pronounced, the Nano did not need discs. In practice, their point is borne out — the drum brakes do a fairly good job of arresting momentum — though for the base version, where the brakes are not assisted, they do need more than an assertive jab.

With simple, independent suspension on all four corners the Nano rides well, dealing reasonably well with road imperfections, though it could have been softer; many may consider it a tad hard. The springing has been kept a little firm deliberately as this set-up has been designed (keeping in mind potentially tricky handling), to provide safe terminal understeer. Getting the spring rates right called for some compromises and deciding on the most appropriate was a tricky exercise.

Yet, despite all the compromises and challenges of making a car to cost, the Nano was seen as a fine product, a point of view shared by several of India's and the world's leading motoring hacks and magazines. True, the Indian press did get hyperbolic, but there were some objective comments too, specifically from the international press.

Yet no one ever imagined the Tata Nano would not succeed in its main market — India. But with the price getting familiar with even conventional cars at the bottom of the Indian marketplace, did the main USP of the Nano get too undermined? Was that the main reason the Nano failed? Or were there other reasons?

Chapter X
Looking Back, Looking Ahead

The Indian automotive industry has come a long way since the country became a sovereign and independent nation. Against sales of 14,688 cars in 1950, Indians bought as many as 2.95 million cars, SUVs and multi-utility vehicles in 2013 — a growth of 180 times, in a little over six decades. And this is just the beginning. In 2009, just 18 people in a thousand had a car in India, as compared to the developed countries' average of well over 500 cars per thousand inhabitants. Maybe India will never get to that. And for the sake of the environment, let us hope it does not. Yet, there is a long way to go in fulfilling the human desire for mobility, travel, freedom and (very importantly, in India) status. Thus, every Indian should have the right and the opportunity to experience the satisfaction of owning and using an automobile.

In 2013, India was the sixth largest producer of automobiles (four-wheelers that is, including cars, SUVs, mini-vans, utility and commercial vehicles), with almost 3.88 million made, behind China, the USA, Japan, Germany and South Korea. By 2020, India is expected to get to the number three spot. Of course, with a faltering economy that may not happen by then, but sooner or later, India is going to get there. And eventually to the number two spot. Incidentally, India is, and has been for many years, the world's second largest motorised two-wheeler market with sales in excess of 12 million, second only to China. But we will come to the future later.

Choices, choices

The past is what this book has been about thus far. And it is worth looking back and understanding how things happened and evolved. The first three decades of India's nascent car industry happened mostly under Governmental regulation, as India chose the planned economy route towards growth. Given the circumstances and the international environment then, it must have seemed the best choice in order to make the most of limited resources.

It was the choice several other democratic nations took as well — Japan and South Korea included. In Japan, though, the automobile industry was a priority sector, initially for commercial vehicles and two-wheelers, but soon thereafter, for cars too. For two-wheelers, governmental regulation was non-existent and a free-for-all followed with just four extremely strong manufacturers surviving and dominating the world even today.

South Korea too, followed the planned economy route. The first step was providing the necessary protection, followed by a Japanese-style Government-backed export-led growth, encouraging a select lot of five car-makers who were mostly part of the chaebol conglomerate system (many companies clustered around one family-owned parent company), similar to the oligopolistic system the Indian economy evolved into.

A more recent example is the Chinese auto industry, which emerged as the world's largest just a few years ago — an excellent example of planned growth with intensive state intervention.

So a planned economy was not the problem of the fledgling Indian car industry. The problem was the very low priority the car industry was accorded by the Indian Government, and the 'planning' that followed as a consequence of 'socialistic' thinking in the form of Licence Raj, which emasculated the industry's growth potential whilst benefitting certain oligopolistic groups. That remained the status for the best part of two decades after India's independence.

True, there were several attempts at a rethink and several proposals and suggestions for a people's car, but it was only in the Fourth Five-Year Plan period that the car industry was accorded some importance and that too, to essentially serve nepotistic ambitions, when Sanjay Gandhi received the licence to make his idea of a people's car. And as we now know, India's first attempt at a "car for the people, by the people and of the people" was an utter disaster. Waiving aside Sanjay Gandhi's abilities to design and develop an indigenous car, what remains a big question mark is whether the car was appropriate for India. The Standard Herald had proved that Indians did not care for two-door cars (as most Indian car buyers preferred four-door cars) so Sanjay's Maruti, with its two-doors, in all probability, would not have worked in the Indian marketplace even if he had managed to sort out the car in engineering terms.

So, after more than three decades of Nehruvian thinking and Fabian socialism, India's car industry reflected the state of the country's economy in more ways than one.

Of filial loyalty and liberalisation

The second attempt at India's people's car also came through thanks to filial reasons — Indira Gandhi's desire to rehabilitate son Sanjay's reputation and 'legacy', even though it was couched in policy terms as an attempt at "modernising the automobile industry, effecting more economical use of scarce fuel and ensuring a higher production of vehicles". The approach this time was, of course, more professional, the attitude more thorough and misguided delusions about abilities did not play a role anymore. The car — the Maruti 800 — was for the people and by the (tax-paying) people of India, though it was not of the people of India. With the Maruti 800, the scenario changed. This little 800cc car finally become the engine of growth for the Indian automobile industry, four decades after India's independence.

Yet, even after the successful launch of the Maruti 800, the Government's shortsightedness persisted, with blatant protectionism favouring Maruti Udyog and discouraging others from embarking on alternate people's car projects. But the financial crisis of 1991 changed all that. The government did a complete about turn and opened up the Indian economy, with the automobile industry 'liberalised' fully for the first time. Not because the Government thought that it made good economic sense for the nation, but because the International Monetary Fund, which was bailing India out of its foreign exchange crisis, insisted on a set of rather radical reforms.

So, from a rigidly controlled industry, the Indian car segment became the showpiece of an open economy, with limited restrictions and an almost laissez-faire regime. Which translated into a spate of joint ventures between Indian car-makers or industrialists and multinational auto majors, most of which, as co-operative ventures, failed eventually. The exception was the wholly-owned subsidiary set up by Hyundai Motors, with more than 100 per cent commitment to the market reflected in terms of significant investments, extensive localisation and redesigning of its first product to suit the Indian market preference. Hyundai not only quickly got to the second rank amongst the car-makers in India, but also started making money fairly early on.

Clearly, joint ventures in a free market economy, do not work, at least not in India. It will be interesting to see how they pan out in the long-term in China, where all the multinationals are into joint-ventures.

The other lesson was that designing, or at least tailoring, a product for India was a very important ingredient for success. Both the Tata Indica and Hyundai Santro were good examples of this hypothesis. The Maruti Alto 800 is another case in point.

The third lesson was that a car that had been successful in some other markets may not necessarily succeed in India unless modified to Indian tastes and requirements, which was done to both the WagonR and the Zen Estilo. The Suzuki WagonR's Japanese success or the Renault/Dacia Logan's worldwide success did not translate into significant sales in India.

Of course, the most important lesson was the cascading effect that a reasonably priced people's car could have on the industry and the economy. Till the Maruti 800 came by, the Indian car market grew at a little more than the 'Hindu growth rate' (of 3.5 per cent or so for the Indian GDP), averaging seven per cent between 1950 (when 14,688 cars got sold) and 1980 (with 30,989 cars sold that year). In the 1980s, with the Maruti 800 introduced mid-way through the decade, the growth rate spiralled upwards to 56 per cent per year with as many as 174,633 cars sold in 1990.

Into the 1990s, the growth rate of the car industry stabilised at 40 per cent per year, and then at a still impressive gallop of 34 per cent every year in the first decade of this millennium, to 700,925 cars sold in 2000 and over 2.37 million cars sold in 2010. And driving these amazing growths were the 'millionaire' cars: the Maruti 800, the Hyundai Santro, the Tata Indica and the Maruti Alto, amongst others. The Maruti Van/Omni is another 'millionaire' vehicle, having crossed the 1.5 million mark in 2012, but it was not considered for this book as it is a van that has been used for passenger purposes and not a 'real' car.

Of course, the first form of governmental 'incentive' to encourage a section of the Indian automobile industry, was the excise differential. Introduced in 2006, this made the small cars (defined as cars shorter than four metres long, with engine capacities of less than 1200cc for petrol-fuelled cars and 1500cc for the diesel-fuelled), more attractive than bigger ones. In the process, this resulted in a spurt in the demand for small cars, accompanied by the necessary industrial environment to make India a global manufacturing base for smaller cars, which, by 2013, represented over two-thirds of India's total car market.

When the market was still young, any inexpensive car with the minimal requirements would have worked — any modern fuel-efficient four-door four-to-five-seat car at Rs 50,000-odd would have sold well in the 1980s. The Minica from Mitsubishi or the 205 from Peugeot, would have done as well as the 800. But as the market matured so did the specificity of the kind of car people wanted: first-time car buyers wanted more space and along came the Tata Indica and Hyundai Santro. People wanted more style and there was the restyled Santro, allowing the stylists to reject cars like the WagonR. Clearly, a purely Indian car "for the people, by the people, of the people" should have worked. The Tata Indica did. Why not the Tata Nano?

At the time of its launch, the Tata Nano had all the ingredients of success, and most expected it to set the sales charts afire, not only in India but across the globe. So why did the Nano fail? For the matter, has it really failed or is it too early to write it off?

After buyer Ashok Raghunath Vichare received the keys to the first Tata Nano on 17 July 2009, not even 250,000 cars found buyers till end 2013, four and a half years after its launch. Yet the planned capacity at the Tata Motors plant is 250,000 Nanos *per* year....

At the beginning, it took Tata Motors a while to get manufacturing going as it was making the cars at its 'interim' facilities in Pantnagar, in Uttarakhand, but by June 2010, the Nano factory at Sanand, in Gujarat, was ready and cars began rolling out. Yet, for the calendar year 2010, Tata Motors managed to shift just 59,576 Nanos. By 2011, the numbers had got better, with 70,784 Nanos sold in India; that climbed to 76,747 by 2012. However, by late 2012, sales of the Nano began sliding. And for the calendar year 2013, just 18,447 cars were sold in India, as compared to 76,747 for the same period the earlier year — a crash of 76 per cent. Many have written the Nano off. What went wrong?

Several factors seem to have contributed to the Nano's lack of success. The first big problem was the series of fires when several Nanos combusted thanks to the short circuit on the switch in the indicator stalk that controlled the headlamps and windscreen wiper. That, for sure, frightened away many customers even after the issue had been addressed. There have been no further incidents since 2011.

Industry veteran, B V R Subbu, believes that when very difficult price targets had been set (building the car with a Rs 75,000 out-of-the-door price point), some compromises in quality are bound to happen; that something is going to give eventually. This point of view has also been expressed by several engineers within Tata Motors, who say that meeting the price and component cost targets had become such an obsession in the company that quality aspects took a back seat in many instances.

That is also reflected in the long-term quality issues. Mumbai-based garage owner and car enthusiast, Arvind Swadi, bought an early Nano, in February 2010, for his wife's daily commute to work. Initially, Arvind and his wife, Dusa, were both happy with the Nano's convenience and practicality, but a few months into the ownership brought out several quality irritants that have been bothering them. Arvind points out, "The car is rusting at several points. The suspension needed to be changed as there was a rattle. The indicator lever needed to be repaired and the headlamps had water seeping in. At the same time, the Tata Motors service has been very responsive and the company has repaired or replaced all faulty parts."

The so-called virtues of convenience and practicality of the Nano, have also come in for criticism, especially since the launch of the Maruti Alto 800. One of the first shortcomings of the Nano (pointed out at the time of its launch), was the lack of a rear 'hatch' or boot opening. In fact, at the launch, this aspect was acknowledged by Ratan Tata himself and Tata spokespeople mentioned they would be looking at a redesign. More than five years later, the Nano still has a fixed rear windscreen.

The first test drives showed the car needed power steering, at least as an option. While the Nano could do without power steering, many felt it would help for tight parking. Yet, optional power steering was one of many obvious improvements the car did not have the benefit of till end-2013.

The interiors too, came up for criticism early on and with the decidedly funkier and classier fascias of the Hyundai Eon, and more recently, the Maruti Alto 800, the Nano needed an improved dashboard — which it finally received in June 2013. In fact, soon after the car's launch, there was talk of several future developments, other than the rear-window hatch, optional power steering and better interiors — a diesel engine variant, an optional five-speed gearbox, a CVT (continuously variable) type automatic transmission option, and a more powerful three-cylinder engine version with disc brakes (as a by-product of the European version Nano).

Tata unveiled the Nano Europa at the 2009 edition of Geneva. © Gautam Sen

At the 2009 edition of the Geneva Motor Show, Tata Motors showed a markedly handsomer version of the Nano with the Nano Europa moniker, supposedly a preview of the European version with the more powerful three-cylinder engine, developing about 60bhp and sporting disc brakes with ABS (anti-lock braking system to reduce the chances of skidding under emergency braking), power steering and other electronic gadgetries to meet European norms and expectations. Tata executives hinted that this version would also be sold in India. Tata Motors also showcased an all-electric version of the Nano.

Four years down the line, there was no sign of any of these promised features. The rear window hatch is expected in late 2014 and the diesel engine version and the European Nano may have been abandoned altogether. In fact, Tata Motors initially worked on developing the Nano for Europe, but as extensive modifications were called for to meet European crash test and safety norms, and the engine needed to be upgraded to a more powerful three-cylinder derivative, the realisation dawned that developing an all-new car may made better sense (and would cost as much as getting the Indian Nano up to scratch). So the company turned to developing an all-new car for Europe. Showcasing this, Tata Motors unveiled the Pixel concept at the 2011 edition of the Geneva Motor Show. But given the basic Nano is not selling in its primary market, it is rumoured that Tata Motors may have shelved its European project for now.

Clearly, the Tata Motors planning since the Nano came out, has been a mess. The time it took to develop the Nano — six to seven years — was not unimpressive when one considers that it was a clean-sheet, back-to-basics design that started from first principles. But the time the company has taken to iron out or address some of the obvious shortcomings, has been agonizingly long and is a reflection of the hubris, post-Nano, of the engineers and designers within Tata Motors. The fact remains that the Nano was not exactly revolutionary or cutting-edge, yet sections of the press made heroes out of the Tata Motors' engineers, who not only lapped it all up but became undeservedly arrogant in the process.

But more than the product shortcomings, the bigger problems have been in marketing, more specifically the positioning, of the product itself. When Ratan Tata was quoted as wanting to make a one lakh rupee car, the price point became the biggest talking point and eventually, a trap. The world erupted about the price point the day the Nano was unveiled and Ratan Tata said: "The car will be priced at Rs 1 lakh. A promise is a promise!" In

India, it was the *lakhtia* car; in the world it was the US$2500/£1,300/ € 2,000 car — an absolutely unbelievable price for an automobile that was not an apology of a car.

But by the time people had lined up to buy the car, they were paying at least 35 per cent more, as the Rs 100,000 was the ex-factory price of the base version and did not include taxes, etc. And for the buyers (three-quarters of the total buyers), who opted for the higher trim levels with air-conditioning and other options, the ex-showroom price was 60-80 per cent higher —far away from the Rs 100,000 promise that had attracted them to the car in the first place. Since 2009, the price of the Nano has only moved northwards, further away from the Rs 100,000 and steadily closer to its main rivals, the Maruti 800, the Maruti Alto 800, and the Hyundai Eon, into territory where the product falls short on several counts when compared to the competition.

Though the *lakhtia* tag gave the car unprecedented exposure and fame, it has also been a huge constraining factor, both in terms of cost and consequently the quality of the parts and components, as well as in terms of cognitive dissonance for buyers as they put down more than twice that amount, wondering all the while whether the car was worth that much.

The biggest problem though has been in the positioning of the Nano. When it gained fame as the world's cheapest car, it also got stigmatised as cheap, versus affordable. In status conscious India, the gap between cheap and affordable is indeed vast, and an automobile, or any product for that matter, ceases to be desirable the day it is seen to be cheap. Tata Motors, with the attitude that it did not need to promote the car (the reasoning being that the launch hype was enough to sell what everyone seemed to want rather desperately), completely missed out on the image aspect.

In delivering the first car to Ashok Raghunath Vichare, Tata had the perfect first customer — a salaried white-collar middle-class customs officer buying his very first car. By the third or fourth car, Tata ought to have handed over a Nano to a star, a socialite — a person who bought the car because they wanted one, not needed one. Instead, the initial impression of a cheap car turned rapidly into a car for people who could not afford any better. In fact, to many, it has become the car they would *not* have at any cost.

It was in 2012, that popular American TV show host, Jay Leno, ordered and was delivered, a specially done up Nano in the USA — a potential publicity and image opportunity that Tata Motors did not capitalise on. Like Jay Leno, many enthusiasts and Westerners appreciate the Nano's back-to-basics approach, an attribute lost on most Indian car buyers, especially first-time buyers. While most Indian customers are attracted to bells and whistles, Europeans see a virtue in the Gandhian belief of 'less is more'. The Nano's functional simplicity is what got French journalist, Julien Bouissou, to buy a Nano in 2012, a year after he moved to India as the Delhi correspondent of French newspaper *Le Monde*. And the pleasure of ownership and convenience was reflected in an extensive one-page article Bouissou did for *Le Monde*, on why the Tata Nano's sales had slowed down.

Germain Araud, another young Frenchman in New Delhi, working as the Country Director for consultancy firm Altios International, believes: "The Nano has not been marketed properly, since the beginning... The media is to be blamed too, as it has created too much of a negative feeling, after the few 'fire' incidents. Plus, the usual clichés — Tata is not an aspirational brand at all. And a big chunk of people targeted for this car at the beginning — the middle class people moving from a bike to a car — cannot afford to buy this car as it is three to five times more expensive than a 100cc motorcycle."

A happy owner of a Nano since 2010, Araud, who bought an old Royal Enfield Bullet some years back when he was working in Mumbai (riding it back to Montpellier, France, when he headed back after his Indian tour of duty), specifically chose to acquire a Nano despite a larger budget. "My interest in the Nano has not faded a bit since I acquired the car in 2010. I am planning on buying another Nano soon!" So, what does Araud like so much about the Nano? "It is everything you want for a car for use in a city and in general for Indian road conditions thanks to its high ground clearance," explains Araud. "The interior is very spacious for the size of the car and the exterior design is very modern and pleasant to look at. It is amazingly good value for money and the servicing is very cheap. The engine is peppy enough to zip through the crazy Delhi traffic. Overall, I like the idea that the vision of one man has been transformed into reality and has brought out the best of Indian engineering. I'm very saddened by the fact that people in India are not objective enough about this car."

Looking Back, Looking Ahead

And what would Araud like to see improved in the car? "The boot doesn't open from outside and that needs to be addressed. Also it could do with better noise insulation. Plus the suspension is too hard. Also the car should have a more up-market feel to the cabin."

Germain Araud and the love of his life. © Vikram Bohra

With its 2013 version, the Nano has at last addressed the image aspect, presenting a more up-market feel. But several other product-related aspects need to be improved significantly, and the sooner the better. Most importantly, Tata's advertising and marketing of the Nano needs to be rethought. In fact, it is too early to write off the Tata Nano, even if certain industry leaders such as R C Bhargava believe otherwise.

History lessons

The automobile industry is full of case histories of how people's cars have 'bounced' back after initial problems. The Volkswagen Beetle had a very inauspicious start. The earliest Volkswagens were called Kdf-Wagens, or Kraft durch Freude Wagens, German for 'strength through joy cars', a Nazi-inspired movement meant to give the German people a sense of purpose, whereby they could have their Kdf-Wagen by purchasing RM5 stamps every week; once 50 stamps were filled, the car would be allotted. 336,638 Germans signed up. With WWII intervening, not a single person ever received a car.

After WWII, when the Volkswagen factory was offered to Ford, the American company refused to have anything to do with it as it believed the factory and the product were both worthless. After production restarted in 1945, sales were very slow. It took Volkswagen the best part of six years to sell its first 250,000 Beetles. It was only in the mid-1950s, almost two decades after the car was conceived, that the Beetle started selling in significant numbers. At the beginning, most sales were in a Germany recovering from the World War, with considerable appetite for cheap personal transport. Then the American market took to the Beetle and that is when the sales of Volkswagen skyrocketed.

But one must remember that the mainstream American market then was a paradise for humungous chrome-laden barges powered by gas-guzzling six- and eight-cylindered engines. Yet the laughable but lovable Beetle made an amazing marketing breakthrough, thanks to some clever, self-deprecating advertising that remains some of the finest examples of how to turn a perceived disadvantage into a virtue or a politically appropriate advantage.

The Mini too, had a difficult start. With an accelerated development programme, it had several design flaws when it was launched, the most infamous being the 'floating carpet' problem — when driven in the rain, water would get into the car, wetting the carpet and making it float! The editor of *MotorSport* magazine wrote: 'When driving the 'World's Most Exciting Car' I found it to live up to its reputation — part of the excitement being to see which foot got wet first!' Apparently, the late addition of a reinforcing box sill to the outer edge of the body caused this water leak. The only solution was to redesign and retool the floorpan, plus repair 8,000 leaking Minis under warranty!

That was not all — oil leaks that sprayed the clutch plate, wrongly specified synchromesh that made the gears crunch when engaging, exposed plugs and distributors at the front of the engine that had the car spluttering to a stop in the rain and others. The car was seen as a bit too clever for the blue-collar workers who were targeted as the initial customers. Many of them thought the car was way too small for their families despite the space within, equating size to status and preferring to go for bigger, traditionally engineered cars such as Ford's 100E Popular, which, ironically was not only slower, with poorer handling, it was no more roomy than the Mini either.

It is when the Mini went rallying, proving it could run rings around pricier cars, that the beginnings of cult-car-status-stardom began for this brand. That was cemented when Princess Margaret, George Harrison of Beatles fame, and Peter Sellers, started a trend for celebrities opting for the Mini as a funky, trendy, classless car with character.

Other than allowing the development of the sportier Mini Cooper version, various other derivatives came off the Mini base, including an estate, a semi-off-road beach vehicle called the Moke and a convertible eventually. Similarly, Volkswagen also encouraged and worked towards developing a range of models derived from the Beetle. The chassis was made available to coachbuilders such as Hebmuller, Rometsch, Drews and Dannenhaur & Stauss. During WWII, the Jeep-like Kubelwagen and the amphibious Schwimmwagen, had already been developed. Other than supplying powertrains to Porsche for their fledgling sports car project, Volkswagen also developed the Kombi bus and supplied chassis-mechanicals to coachbuilder Karmann to make the Volkswagen Karmann-Ghia sports coupes and convertibles. Soon there were half a dozen different kinds of vehicles from the Beetle family — from a practical two-door four-seater, through 2+2-seater coupes and convertibles to a seven-seater minivan.

The Citroen 2CV too, had a slow start. But the reasons were very different. People wanted it but the car-maker could not make enough, thanks to post-war supply shortages of material, specifically steel. It was only in its seventh year of production that the 2CV crossed the 250,000 cars mark, and it was only in 1957, eight years after start of production, that the 2CV was selling over 100,000 units every year. The press disliked it intensely, with an American motoring journalist proposing that a can opener be offered with the car and British car magazine *Autocar* concluding that the 2CV was the 'work of a designer who has kissed the lash of austerity with almost masochistic fervor'. Appreciation from journalists came much later, with respected British motoring hack, L J K Setright, declaring the 2CV 'the most intelligent application of minimalism ever to succeed as a car'.

Thus, many people's cars through time have had difficult starts. Brilliant advertising made the Beetle the great success story it finally was, the Mini had serious quality and perception issues to surmount, and Citroen just ignored what the press had to say, plugging along its own quirky, individualistic route. Fiat, with its 500, also developed and encouraged a whole host of variants and coach-built specials. Other than developing a convertible version and an estate, an entire range was built by sister company Autobianchi as a more up-market alternative to the Fiat 500 range, and chassis-mechanicals were provided to specialists such as Abarth, Giannini and Vignale to built hotter, sportier versions, or pretty convertibles and sports coupes.

Citroen did the same with the 2CV. A van version went on sale just two years after the 2CV was launched, in 1951. In 1960, Citroen launched the 2CV Sahara, with an engine mounted at the rear driving the rear wheels, retaining the front engine, as a 'four-wheel-drive' vehicle! A more practical semi-off-road vehicle was the Méhari, sporting a washable plastic body, of which almost 145,000 were made from 1968 to 1987.

But the most noteworthy derivative of the 2CV was the Ami. Based on the 2CV platform, but with a slightly more powerful derivative of the engine, the bigger, better equipped and more modern Ami was a product designed to slot in-between the low cost 2CV and Citroen's flagship model, the big and sophisticated Citroen DS, and was thus priced some 35 percent higher than the 2CV. Initial sales was slow, but after a few detailed improvements, sales finally took off and for many years after, the Ami was France's bestselling car, outselling the 2CV handsomely. With a production run of 17 years, the Ami sold over 1.84 million units. Yet, the 2CV outlived (and outsold) it by 12 years.

In fact, the Ami, not unlike the Beetle and the Mini, is an instance of a model being a bestseller despite not being the cheapest. As markets mature, they evolve to a level where the cheapest is not necessarily the bestselling. When the Maruti 800 came in, the Indian market was nascent and more than eager to buy a cheap and cheerful car that was significantly better in many ways to the existing offerings in the marketplace. But as the range of models widened and the consumer matured, the cheapest is no longer necessarily the best option for many. The success of the Maruti Alto and the Swift, both more expensive than the 800, proves that the Indian car market has matured too.

Looking Back, Looking Ahead

So for the Nano to succeed and win back a following, a concerted effort will need to be made to change its image by moving further up-market in terms of versions and derivatives. According to an article in *The Economic Times* of 23 February 2013, Tata Motors has been working on a family of low-cost, composite niche cars to complement the Nano. When the Nano was being developed, Ratan Tata had been keen to explore the possibility of developing a plastic/composite-bodied, low-cost car. But the Nano eventually turned out to be a rather conventional, steel-bodied automobile. Developing body variants from a steel monocoque base is capital intensive, other than calling for extensive engineering. Whereas composite-based niche derivatives (apparently a coupé and a small SUV were under consideration), using the Nano mechanicals, could be developed at significantly lower cost and act as halo products around the Nano, improving image and giving the mechanical aggregates and component volumes needed to amortize costs and investments.

In fact, in a sense, the Nano (as it has been designed), may have completely missed the point. When Ratan Tata wanted to develop a car for the people of India, it may have been worth noting that two-thirds of India lived, and still lives, in rural and semi-rural areas, in conditions where the small-wheeled Nano is just not the right product. A vehicle with bigger wheels and greater suspension travel would have been more appropriate. It is a promising segment that will, in all likelihood, be better addressed by other carmakers, starting with clean-slate designs such as Renault, for instance.

Without doubt, opportunities have been missed and Tata has been too slow to react and change strategies. But it is still not too late and 2014 will be a crucial year for the Nano, a turn-around year for a car that still has a lot going for it. Unless the new management has decided to 'abandon' it.

Crystal-gazing

And if it is crystal-gazing that we are at for the Nano, how about the other 'millionaire' cars of the future? What can we expect as people's cars of tomorrow? From the Maruti stable, the Swift joined the 'millionaire club' in 2014, though that counts for two generations of the car. By the same argument, the WagonR has qualified already, but then, it is over two gens too. The Maruti Alto 800 has the potential to take on the mantle of the 800 and the Alto, becoming another 'millionaire' model in a few years.

From the Hyundai stable, the i10's production figure comfortably tops the million, but as an India-selling 'millionaire' it is still off the mark. But by 2016 or 2017, it could get there. The Eon too, has the potential, but it is a case of another low-cost car that has not quite hit the sweet spot yet. Whereas, the pricier but markedly bigger i10 Grand, on last count, was setting the sales charts afire.

The other so-called 'low cost' cars from the Japanese giants have had, at best, mixed success, with both the Toyota Etios Liva and the Honda Brio selling in modest numbers. In fact, what is more noteworthy is the success the saloon-derivative of the Honda Brio, the Amaze, has had since its launch in India in April 2013, outselling its cheaper brethren two-to-one. In a sense, the Amaze's success reflects that of the Maruti Swift Dzire, with the latter making for a very interesting case study, and perhaps, in some ways, predicting the future. For the month of May 2013, Maruti Suzuki sold 17,265 Dzires, outselling the Maruti Alto as India's bestseller since July 2004 (except for April 2012, when for that one month the Swift outsold the Alto), marking the first instance, since 1985, of a saloon being the bestselling car in India.

What is more, for the calendar year 2013, 198,317 Dzires were bought by Indians, making it the third bestselling car in India after the Alto 800 and the Swift. For the best part of the last three decades, the bestselling cars in India have been small hatchbacks like the Maruti 800, the Alto, the Santro, the Indica, Swift, and so on, convincing industry experts that India, unlike the rest of Asia, was a small hatchback market. But is that really so?

In fact, the very basis and argument for the excise differentiation in India has been on the premise that the country is essentially a market for small cars and that Indian consumers — unlike most of Asia and like southern Europe — having a clear preference for small hatchback cars. Thus the argument that by stimulating and encouraging smaller cars in India, the impetus for viable volume for the domestic market would spin off into sustainable volumes for international markets, most specifically that of Europe. The fact that exports to Europe and other international markets have been on the increase, to the point where over half a million cars were exported from India during April 2011-March 2012 for the very first time, would confirm the Indian Government's first proactive legislation in differential excise has been beneficial for the Indian automobile industry.

Along with the price differential between petrol and diesel, with the latter cheaper, India's car market, in more ways than one, reflects that of southern Europe, in that Indians seem to prefer small, diesel-engine hatchback cars with manual transmissions — which is what most car buyers opt for in countries like France, Italy and Spain. The market composition is also distinctly different from that of most of Asia: in China the bestselling car is the 4.6-metre long saloon, the Volkswagen Lavida (which is somewhat similar in size to a VW Jetta), Japan's bestseller is the 4.5-metre long Toyota Prius, South Koreans prefer the huge 4.8-metre long Hyundai Sonata, in Vietnam, the big and mighty 4.7-metre long Toyota Fortuner is popular and the Thais have a penchant for big pick-up trucks.

It is only in Indonesia, Malaysia and Iran, that the bestsellers are smaller cars — the Indonesians' first choice is the 4.1-metre Toyota Avanza, Malaysians like the Perodua MyVi, which is just 3.7 metres long, and Iranians have a preference for the under four-metre Saipa Pride. Yet almost all the cars selling in Asia are petrol-engined, and significant numbers are automatics.

Given that more than two-thirds of the Indian market is for small under-four-metre cars, five-door hatchbacks with the option of diesel as well as petrol engines, and with mainly manual transmission, most experts and analysts would tend to believe that future investments and Indian bestsellers will be these kinds of cars. Yet the success of the Swift Dzire and the Honda Amaze (and the success the similar, trendsetting Tata Indigo CS had in the past), indicates that, given a choice, Indians may prefer saloons over hatchbacks. Or is that so?

Well, the success of the Skoda Octavia in the past also indicates that at the prestige end of the market too, a hatchback can be more than acceptable if the car *looks* like a saloon. Thus, the success of the Dzire, the Amaze and the Indigo CS, probably indicates a very interesting trend — that despite being less practical, people would prefer a car that *looks* like a saloon over a hatchback. And that is why car-makers such as Datsun, Ford, Hyundai and others are all planning to make sub-four-metre saloons. In which case, the market could rapidly skew away from hatchbacks to saloons and the excise differential would at best create strange 'mutants' in the form of the Dzire and the Amaze, which would have very limited appeal beyond the frontiers of India. The strangest of this new breed, is the Mahindra Verito Vibe, the shortened saloon version of the Verito, which does not have the practicality of a hatchback, despite looking like one.

Tata's little booted car, the Indigo CS. Courtesy Tata Motors

Thus, for the longer term, the excise differential (the way it is structured currently), may not be all that beneficial as more and more Indian consumers move away from cars that are 'exportable,' to cars that will have very limited appeal in most other markets of the world. What could be the solution to addressing this possibility?

Encouraging the manufacturing of smaller cars makes a lot of sense as most developed markets are downsizing and will keep doing so; plus, smaller cars consume less fuel, are cleaner and as cities get more and more congested, address the increasing issue of size and road encumbrance. So a telescopic taxation system, with higher excise for bigger cars but applied progressively as per size, may be the best solution for the future, instead of creating an inflexion point at exactly four metres, making the smaller cars less taxed, with bigger cars at progressively greater rates of taxation. Plus, this telescopic rate of taxes could be coupled with another progressive system or rate of penalization for emission figures: cleaner cars, whatever the technology that they use, should be taxed lower than cars emitting greater quantity of greenhouse gases. So, instead of providing and spending substantial moneys in supporting a technology — such as electric cars — the Government of India should encourage all technologies that reduce emissions and consumption of conventional fossil fuels.

If such a system is followed, than some 'natural' small saloons might come into being — models like a 4.1-metre long Honda Amaze or a 4.2-metre long Swift Dzire, cars which will also find buyers in other markets of the world, instead of being oddities specific to the Indian market space. If car-makers resort to cost-based pricing, these saloons should not be more expensive than their hatchback counterparts, as it does not cost more to make a saloon that shares most parts with a hatchback. And if saloons' prices are close to that of their hatchback counterparts, this writer believes that saloons will outsell hatchbacks significantly in India.

This writer also believes the Indian consumer prefers saloons and bigger cars and India, like most of Asia, would have been a big car market if bigger cars were more affordable. Also, the Indian consumer actually prefers the three-box saloon *looks* to that of the two-box hatchback design, and if a three-box-looking hatchback (like the Skoda Octavia), is made, the Indian consumer would opt for it as much as a booted saloon (if not more so, as the former would be more practical).

So what are the implications of all this for the 'millionaire' cars? The fact is that the next generation of millionaire cars could very well be small saloons or hatchbacks-that-look-like-saloons. Some out-of-the-box designs could very well be the future of motoring for India's millions; the next 'millionaire' could well be an automobile that is not quite a car but a quadricycle.

In May 2013, the Government of India decided to set up an inter-ministerial group, excluding the manufacturers, to come up with a consensus for safety and emission norms for quadricycles. For several decades now there has been an effort (by the automobile industry and Western society in general), as part of a global initiative to improve urban mobility in the form of small, space-efficient vehicles that are low in fuel consumption and emission and with limited road encumbrances, deemed more appropriate for in-city use.

Given the safety issues concerning two-wheelers and reluctance in their usage by sections of society, a safer and more practical alternative has been seen in the vehicles defined as quadricycles — basically four-wheeled motorcycles which have the virtues of a two-wheeler (small size and limited road encumbrance, less fuel consumption and lesser pollution), yet the greater safety, convenience and comfort of a car. The market potential for such vehicles is, therefore, tremendous and any move from a two/three-wheeler to this segment is not only natural but also the next logical step.

To draw a parallel, the scenario pertaining to Kei cars in post-war Japan is a case in point; this was a time when successful motorcycle manufacturers like Suzuki and Honda moved from making bikes to small cars that could use the bike engines they were already manufacturing. Three-wheeler manufacturer Mazda, as well as Daihatsu, followed.

In the Indian context, the idea of using the three-wheeler market as an entry point into the car industry could be revolutionary, specifically for motorcycle manufacturers such as Bajaj Auto, Hero MotoCorp and TVS Motor Company. From a marketing stance, Rajiv Bajaj has played his cards shrewdly. Instead of competing directly with the likes of Tata Motors and Maruti Suzuki in the small car segment, Bajaj Auto has opted to target the present users of its three-wheelers, a market they dominate. If the price of the Bajaj RE60 is competitive, it could even persuade rickshaw buyers to shift to this four-wheeled quaadricycle, which would provide a better commuting experience for those who cannot afford to pay the tariff of a conventional taxi.

It is also possible to design a vehicle that, with the same base, could serve several market possibilities. It should therefore be possible to design a small car to meet quadricycle norms and use the same base to make a slightly heavier and more powerful car to serve as a small city car which could rival the likes of the Tata Nano, the Maruti Alto 800, or for the matter, the Smart in Europe. And even if the more sophisticated version manages to compete against the Nanos and the Altos of the Indian marketplace, the more basic, affordable versions could well be the *lakhtia* vehicle of the future — the Rs 100,000-automobile that would be the eventual bridge between motorised two-wheelers and cars.

So will the future of the 'millionaire' cars be quadricycle-type automobiles or short-butted saloons, or will a mini SUV rule India? Will Maruti, Hyundai and Tata continue to dominate or will Bajaj come by and mess up the pecking order? Or will there be another upstart, another agent of disruptive technology, that will turn the order upside down?

Looking Back, Looking Ahead

Despite a history of over a century, the automobile industry — which time and again has been written off as a 'sunset' industry — has had some strange twists and turns, ups and downs, ins and outs, as the rapid economic growth of emerging markets has changed the dynamics of this industry altogether. And even if the order of the top ten car-makers globally has not changed all that much, down the line there has been a whole lot of changes in the last couple of decades.

Till as recently as 1998, of the top 36 automobile manufacturers globally, barely four Chinese (essentially commercial vehicle) brands, had just about crept in. By 2011, barely 13 years later, of the top 50 four-wheeler manufacturers in the world, 20 were Chinese car and truck makers. And every year these brands and companies are gaining rank by leaps and bounds. Yes, there are three Indian auto-makers on the list too —Tata Motors, Mahindra & Mahindra and Ashok Leyland — and they too, have been steadily climbing the charts. But like the Chinese, the potential remains for more Indians joining the fray as India becomes an important market and manufacturing base for automobiles.

Of course, it calls for inventiveness, aggression, sheer guts and chutzpah... but whether upstart or established, car-makers will need a host of innovative 'millionaire' automobiles to get onto and stay, on the list of the best of the best.

Glossary

A-pillar: Roof support each side of the windscreen
Aerodynamics: Wind resistance, the more streamlined a car is, the more aerodynamic it is
Aftermarket: Accessories and components that one can purchase and fit on the car after delivery, not from the manufacturer
Air cooled engine: An engine that is cooled by ambient air passing over the engine fins and head
Air dam: Device to reduce or redirect airflow
Aluminium block: Engine cylinder block cast in aluminium
Anti-roll bar: Transverse rod between left and right wheels either at the front or the rear axle to reduce body roll
Axle: A rod or spindle passing through the centre of a wheel

B-pillar: The roof and door frame support aft of the driver
Bench seat: Single full width seat
bhp: brake horse power, and is the actual or useful horsepower of an engine, usually determined from the force exerted on a friction brake or dynamometer connected to the drive shaft
Boot: The trunk of an automobile or the vehicle's main storage, luggage, or cargo compartment
Bore: Diameter of the circular chamber cut into the cylinder block
Bucket seat: Individual seats in a car with added support on the sides to secure the driver or passenger
Bulkhead: Panel separating the engine from the passenger compartment, also called a firewall
Bumper: A horizontal baar or panel fixed in the front or rear of a vehicle to reduce impact damage

C-pillar: Side pillars to the rear of the second row of seats supporting the roof
Cabriolet: An open top car, also called a convertible
Camshaft: A shaft with one or more cams attached to it, spcificaally one operating the valves in an internal combustion engine
Capacity: The size of an engine reflecting the volume displaced by very piston
Carburettor: Device in an internal combustion engine for mixing air with a fine spray of liquid fuel

Glossary

cc: cubic capacity – the of an engine measured in cubic centimeters
Chain drive: The transmission of power via a chain passing through two sprockets
Chassis: The base frame of a motor vehicle or other wheeled conveyance
Clutch: Device for engaging and disengaging two working parts of a shaft or of a shaft and a driving mechanism, used to transmit power from engine to gearbox
Coachbuilt: Specially built bodies built by manufacturer of bodies for carriages or automobiles
Coil spring: A helical spring made from metal wire or a metal band used for vehicle suspension
Coupé: A two-door car with a hardtop, with or without two extra seats in addition to the two at front
Cylinder: The hollow part of the engine block where the pistons travel, usually circular in cross-section
Cylinder head: The closed, often detachable, end of a cylinder in an internal-combustion engine, which carries the valves and sparkplugs, and in some cases, the camshafts

Diesel engine: Internal combustion engine which burns diesel oil instead of petrol
Differential: Gears that permit the wheels to turn at different speeds when the car is turning
Disc brake: System of braking where friction pads are pressed against a flat metal surface
Displacement: Same as engine capacity or size of the engine
Driveshaft: A rotating shaft that transmits mechanical power from the differential to the wheels
Drivetrain: The entire engine, gearbox and differential aspects of an automobile, also sometimes referred to as powertrain
Drum brake: Braking system where the friction surface is radially within the metal cylinder
Ergonomic: Layout of controls to ease human-machine interface
Exhaust: Device through which vaporous waste material is released from the an engine

Fascia: A car's dashboard or instrument panel
Fastback: Body style of a car with steeply sloping rear section
Fender: Wing or mudguard of a car
Fibreglass: Covering material made of glass fibres in resin
Five-door hatchback: A car with four lateral doors, plus one at the rear that open upwards and gives access to the inside of the car, including the boot

GLOSSARY

Flat-twin/flat-four: Engine layout where either the two or four cylinders are flat and horizontally opposed to each other
Floorpan: Structural floor of the car and may be a part of the chassis too
Four-stroke: Engine based on the Otto cycle requiring four piston strokes for each power stroke
Fuel injection: Direct introduction of fuel under pressure into the combustion units of an internal combustion engine

Gearbox: A set of gears with its casing that transmits power from the engine to the driveshafts or the propeller shaft
Grille: An opening with several slits in the body of a vehicle to allow air to enter to the radiator and engine area
Ground clearance: The distance between the lowest point of the underside of a car and the ground

Half-shaft: Shorter axles connecting the transmission to the wheels
Handbrake: Brake operated manually by the driver, when the vehicle is static
Hatchback: A car with a door (or tailgate) across the full width at the back end that opens upward

Ignition: The process of starting the combustion of fuel in the cylinders of an internal combustion engine
Independent suspension: Automobile suspension system that allows each wheel on the same axle to move vertically and independently of the others
Internal combustion engine: An engine, such as an automotive gasoline piston engine or a diesel, in which fuel is burned within the engine proper rather than externally (like in the case of a steam engine)

Kerb weight: total weight of a vehicle with standard equipment, all necessary operating consumables and full tank of fuel, but without passengers or luggage
Km/h: A measure of speed that gives the number of kilometers covered in an hour
kpl: A measure of fuel consumption giving the number of kilometers covered with one litre of fuel

Ladder frame: A type of chassis that consist of two long rails, located by cross members on which the body of the vehicle rests
Leaf spring: A method of suspension consisting of springs made of a number of strips of metal curved slightly upward and clamped together one above the other

GLOSSARY

MacPherson strut: A type of car suspension system which uses the axis of a telescopic damper as the upper steering pivot
Mid-engine: A layout where the engine is located ahead of the rear axle line
Monocoque: Vehicle structure in which the chassis is integral with the body, with the body acting as the structural member

OHC: Overhead camshaft, where the camshaft is situated above the cylinder head
OHV: Overhead valve, where the camshaft is placed within the cylinder block and uses pushrods to actuate the rocker arms
Panhard rod: A component of a car suspension system that provides lateral location of the axle that was originally invented by French carmaker Panhard in the early 20th century
Piston: The moving component that is contained by the cylinder in an engine and which moves up and down transmitting the energy created by combustion

Rack and pinion: System of gearing used in steering box with a toothed rail driven laterally by a pinion on the end of the steering column
Radiator: A cooling device through which water or other fluids circulate as a coolant
Rear engine: A layout where the engine of the car is located at the rear, behind the rear axle line
Rigid axle: A simple non-independent suspension consisting of a rigid transverse member with wheel hubs solidly bolted to it
Roll bar: A metal bar running up the sides and across the top of a vehicle, strengthening its frame and protecting the occupants should the vehicle roll over or turn turtle
Rolling chassis: The frame of a car plus its running gear
rpm: revolutions per minute measuring the crankshafts rotating speed
Running gear: The moving parts of an automobile, specifically the wheels, steering, and the suspension system

Saloon: A passenger car in a three-box configuration with A, B and C-pillars supporting a covered roof and a boot behind (the American term being sedan)
Semi-independent suspension: A rear axle design with a torsionally flexible cross-member, the wheels being located on trailing links
Separate chassis: Where the body and chassis are separate components, bolted together
Shock absorber: A device for absorbing jolts and vibrations

Glossary

Stroke: In a reciprocating engine the cycle of reciprocation consists of two opposite motions, in one direction, followed by a motion back in the opposite direction. Each of these is called a stroke and the term is also used to mean the length of the stroke of an engine
Supercharger: A device that increases the pressure of the fuel-air mixture in an internal combustion engine, used in order to achieve greater efficiency

Three-door hatchback: A car with two lateral doors, plus one at the rear that open upwards and gives access to the inside of the car, including the boot
Torque: A twisting force that reflects the power output of an engine at a specific point in time
Transmission: General terms for clutch, gearbox, propeller shaft and differential
Transverse engine: When the engine is placed so that the crankshaft lies parallel to the axle
Turbocharger: A supercharger driven by a turbine powered by the engine's exhaust gases
Two-stroke: A type of internal combustion engine which completes a power cycle in only one crankshaft revolution and with two strokes, or up and down movements, of the piston in comparison to a four-stroke engine's four strokes to do so

Valve: Device used for regulating the flow of liquid or gas

Wheelbase: The distance between the front and rear axles of a vehicle

Bibliography

Adler, Dennis. *Mercedes-Benz* (Motorbooks, MBI Publishing Company, 2008)

Bhargava, R C & Seetha. *The Maruti Story: How a Public Sector Company Put India on Wheels* (Collins Business, 2010)

Buckley, Martin. *The Car: A Book of Classic Motors* (Southwater, 1997)

Chacko, Philip; Noronha, Christabelle; Agrawal, Sujata. *Small Wonder: The Making of the Nano* (Westland Ltd, 2010)

Cheetham, Craig. Classic Cars: *Legendary Automobiles Design and Development* (Metro Books, 2008)

Cheetham, Craig, General Editor. *American Cars: The Automobiles that Made America* (Amber Books, 2004)

De La Rive Box, Rob. *The Complete Encyclopedia of Saloon Cars 1945-1975* (Rebo Publishers, 1999)

Design Museum. *Fifty Cars That Changed the World* (Conran Octopus)

Ford, Henry & Crowther, Samuel. *My Life and Work* (Doubleday, Page & Co)

Freiberg, Kevin and Jackie. *Nanovation: How a Little Car Can Teach the World to Think Big* (Portfolio, Penguin Group, 2010)

Gay, Jean-Marc & Semonsut, Alexandre. *La Peugeot 205 de mon père* (E.T.A.I. Editions)

Ingrassia, Paul. *Engines of Change: A History of the American Dream in Fifteen Cars* (Simon & Schuster, 2012)

Mehta, Vinod. *The Sanjay Story* (HarperCollins Publishers India, 1979 & 2012)

Meurie, Benoit. *La Dyane de mon père* (E.T.A.I. Editions)

Pignacca, Brizio. *Fiat 1100* (Giorgio Nada Editore, 1992)

Pressnell, Jon. *Mini: The Definitive History* (Haynes Publishing)

Pressnell, Jon. *Morris: The Cars and the Company* (Haynes Publishing)

Reynolds, John. *André Citroen: The Man and the Motor Cars* (Sutton Publishing, 1996)

Robson, Graham. *The Cars of BMC* (Motor Racing Publications Ltd, 1987)

Robson, Graham. *The Book of the Standard Motor Company* (Veloce Publishing Ltd)

Sara, Bernard. *2CV l'auto aux mille visages* (E.T.A.I. Editions)

Sparrow, Andrea & David. *Fiat & Abarth 500 & 600: Colour Family Album* (Veloce Publishing Plc, 1998)

Acknowledgements

Of course, this book would never have been possible without the huge help I have had from many friends, colleagues (present and past), mentors and well-wishers. A big 'Thank You' (in alphabetic order) to:

Ashok Advani, Wasif Ahmad, V Anand, Germain Araud, Behram Ardeshir, Sunil Bajaj, Makarand Baokar, Silvia Baruffaldi, Sandeep Bhambra, R C Bhargava, Karl Bhote, Jai Bolar, Julien Bouissou, Anne Castagnos-Sen, Vinod Chacko, Serge Cordey, Robert Cumberford, Adil Jal Darukhanawala, Christian Descombes, Atul Deshmukh, Erica deSouza, Anil Dharker, Tutu Dhawan, Maitreya Doshi, Anu Dutt, Richard Feast, Kartik Ganesh, Raphael Galdos del Carpio, Shohini Ghosh, Frederic Guihal, Sam Hardy, Ravi Kant, Dr V Krishnamurthy, Prem Kumar, Ravi Kumar, Alexandra Legendre, Nadia Lewis, Abhishek Mahapatra, Chandralekha Maitra, Stephane Meunier, Rajiv Mitra, Swarup Nanda, Marc Nassif, Verena Polacek, Bhisham Rai, I V Rao, Debasis Ray, Bob Rupani, Cedric Sabine, Amit Sen, Sugato Sen, Arthur Serrao, Deeptie Sethi, Ravi Sharma, Rahul Singh, Anne-Laure Sion, Dr V Sumantran, Arvind Swadi, Ratan Tata, Harit Trivedi and Suchita Vemuri.

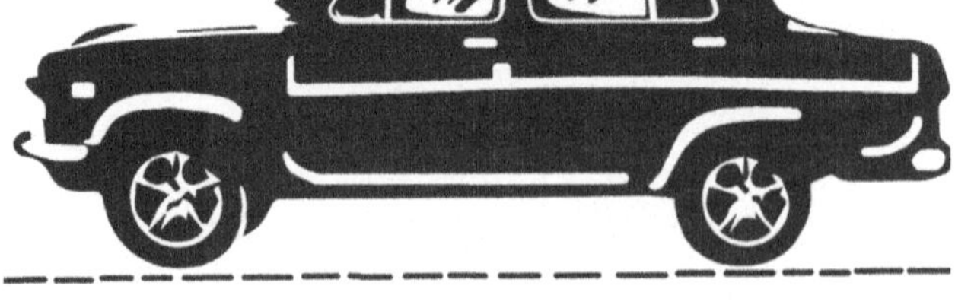

Made in the USA
Monee, IL
08 November 2026

68085078R00152